Data Governance for Everyone

ISBN: 978-0-6452212-0-6

Disclaimer

The material in this publication is of the nature of general comment only, and does not represent professional advice. It is not intended to provide specific guidance for particular circumstances and it should not be relied on as the basis for any decision to take action or not take action on any matter which it covers.

Readers should obtain professional advice where appropriate, before making any such decision.

To the maximum extent permitted by law, the author and publisher disclaim all responsibility and liability to any person, arising directly or indirectly from any person taking or not taking action based on the information in this publication.

To my parents, John and Carole
whose hard work and foresight
created an environment rich in areas
to learn about - one of which was data.

Data Governance for Everyone

A Practical Guide to Unlocking the Value in Digital Transformation and the Fourth Industrial Revolution

Mark Burnard

Table of Contents

INTRODUCTION

Your Key to Unlocking Value in the Fourth Industrial Revolution

The World Economic Forum estimates that the next wave of innovation — based on artificial intelligence and machine learning (or AI/ML for short) — will add over 16 trillion dollars of value to the global economy in the next 10 years.

As I explain in a later chapter, this incredible shake-up has been termed the Fourth Industrial Revolution because of the impact it will have not just on our jobs, but on every aspect of our lives. There will be significant new challenges in areas such as data ethics, AI ethics and of course, job disruption; however, as with the three previous productivity revolutions, the overall impact will be incredibly positive, even for developing nations.

While the raw materials of the First Industrial Revolution were fuels in the form of coal and oil, the raw material of the Fourth Industrial Revolution is data. And as with those other fuels, the *quality* of the raw material critically affects our ability to use it safely and sustainably.

And so we come to the challenge — and the opportunity — of data governance. I've called this book "Data Governance for Everyone" because it's aimed squarely at the non-technical person who, in order to participate more effectively and more fully in the information economy, needs to level up their ability to think critically about data, the role of data, the lifecycle of data, the many pitfalls of managing data poorly, and the incredible potential benefits of using data well. This book will also be incredibly valuable to seasoned data professionals who want to uplift their data governance literacy and provides detailed and solid guidance for the implementation of a data governance initiative in any organisation.

Welcome to the revolution!

SECTION 1

The What and Why of Data Governance

CHAPTER 1

The Critical Importance of Data Governance

The biggest lawsuit in Australian history...

In 2015, the Australian Taxation Office (ATO) used a data matching approach to identify people whom it believed had claimed too much money in government benefits. Over 900,000 letters were sent to 600,000 Australian households, claiming overpayments in many cases in the thousands of dollars (the average amount that was claimed to be owed by the recipients to the government was $2,546.21). These letters were sent to people who were mostly living paycheck to paycheck or surviving on welfare.

This was an incredible shock to the people who received these notices; the initiative became known as “robodebt”, after the automated “robotic” system that generated these distressing notices. To make matters worse, following existing government processes, many of these debts were then passed to debt collection firms who were often very aggressive in pursuing the debts. The result was incredible stress and heartache for those who were targeted. In fact, if you do an internet search for “robodebt suicides”, you’ll uncover one of the most tragic outcomes of that misguided government initiative.

The ATO has since paid out over $1.2 billion to settle a class action lawsuit brought against it on the basis that the debts were incorrectly calculated. In most cases, the money was never even owed, or much smaller amounts were owed than the ATO had claimed. This was, and remains, the most expensive class action lawsuit in Australian history.

And the root cause of this tragedy was poor data governance. Two datasets from two government departments, wrongly combined and interpreted due to an incorrect understanding of the business rules. These incorrect rules were fed into an algorithm, which was then used to drive the robodebt system. In this particular case, the error was that the reported annual income was averaged over the 26 fortnights in a year, and this figure was used to determine overpayments in any given fortnight. Basic data governance — in this case, what we will call Data Usability — would have revealed the fact that the business rules should not have been applied this way because the rules under which payments were made were not calculated this way. But somehow, between the two government departments whose data were combined as part of the calculations, this simple business rule was overlooked. This constituted a basic failure in data governance, which ultimately cost the government over a billion dollars and resulted in unnecessary stress for hundreds of thousands of Australian families.

...was a failure of Data Governance.

Data Governance refers to a system of processes, checks, and balances that provide the "safety rails" around how an organisation uses and shares its data. It includes capturing, curating, updating, and making available the knowledge that is needed to use data effectively and safely — i.e. Data Usability (as mentioned above). It also includes data security and availability (who can and can't access the data), data lineage (where the data came from and how it has been processed on the way to where it is now) and, of course, data quality. All of these we will look at in detail throughout this book.

The failure of this Australian government department was a failure to capture, curate, and share the necessary business rules around the calculations that were used to make payments to recipients. This resulted in a different government department going off and spending many months and many hundreds of developer hours constructing a solution that was based on the wrong business rules (I can't tell you how many times I've seen this happen in organisations large and small). The result was horrendous trauma for many Australian families who were on the receiving end of the results of that data governance failure.

And so was...

Poor data governance can have very expensive results in other ways too.

In September of 1999, scientists and astronomers on both sides of the Atlantic watched in anticipation as the space probe they had spent 5 years working on approached Mars after a 10-month journey through space, following its launch from Cape Canaveral Air Force Station's SLC-17A launchpad. But something was wrong with the trajectory. Worry turned into gasps of horror as the elite group of scientists realised that rather than heading into orbit around the Red Planet, the Mars Climate Orbiter was on a direct course to smash straight into the surface. Over three hundred million dollars of high technology space vehicle erupted in a cloud of red dust. The root cause was later identified as a failure in data governance, in the way that data was shared between NASA and the European Space Agency. One team was using the metric unit Newtons in thrust calculations; the other team was using the Imperial unit Pounds. Basic metadata management, part of data governance, would have prevented this very expensive and embarrassing disaster.

The Big Picture

Sadly, the mistakes made by the Australian Taxation Office and by rocket scientists in both Europe and the USA are not isolated or unusual incidents. According to one study, poor data governance costs Australian companies over $7.8 billion per year. That's $7.8 billion missing from the bottom lines of local companies and value missing from the economy. Put another way, that represents a 4% drain on Australia's $2 trillion economy. In the next chapter, we'll look at similar data from the USA which indicates a cost of US$3.1 trillion from poor data management, a 16% drain on their US$18.7 trillion economy (2016 figures).

Even more importantly, as the value of data increases exponentially — with the rise of artificial intelligence, machine learning and process automation — that squandered value will increase exponentially also. In fact, as we will discover, companies that do not have their data governance "house" in order when dealing with today's data are in no way ready for digital transformation and will be at a significant competitive disadvantage in the data economy that is coming.

In this book, I set out to highlight the value of data — both in the current economy and in the economy that's coming — and to highlight the fundamental and foundational role that data governance plays in your organisation's ability to unlock and harness that value.

We're going to look at what happens when things go wrong in data governance — both in terms of damages and fines as well as in terms of lost opportunities — and also at what happens when things go right!

We will then take a deep dive into what data governance is, what kinds of data we need to govern, how we can govern it, and how to roll out a program of data governance across an organisation.

CHAPTER 2

The High Cost of Poor Data Governance

Banks Behaving Badly

Tropical palms and coconut trees sway in the breeze above the streets of Angeles City in the Province of Pampanga, northwest of the Philippines' capital. But the idyllic scenery hides a dark secret: historically, it has been a major sex tourism hub, frequented both in person and online by pedophiles from all over the world.

In September 2020, Westpac Bank was implicated in enabling monetary transfers between its customers in Australia and parties to child exploitation in the Philippines. In fact, the bank was slammed with the largest fine in Australian corporate history: $1.3 billion. The fine was levied because of over 23 million breaches of the law relating to preserving data about suspicious transactions and passing on such data to the appropriate authorities (in this case, AUSTRAC).

Westpac had been warned specifically by AUSTRAC in 2013, and again in 2016, that they were falling short of their data collection and reporting obligations. In hindsight, it is staggering that known broken processes were allowed to continue unaddressed for almost 7 years, despite warnings from both internal and external parties that something was wrong. In fact, the courts specifically called out "indifference by senior management" as part of the problem. The willingness of senior management to turn a blind eye, to allow inertia to prevail, or perhaps to just bury their heads and say "not my problem", ultimately cost the jobs of CEO Brian Hartzer, Chairman Lindsay Maxsted, and a number of senior managers. In fact, many were surprised the fine was not higher, given that the penalty *per offense* could have been over $20 million — exposing the bank to a potential penalty of over $400 trillion!

The reputational damage may have been worse. Having your name plastered all over the media for weeks and months in association with child exploitation is not a good look for a bank.

These failures occurred at the intersection of data and business processes and, specifically, the legal obligations of the bank to manage its data and to report certain kinds of data in specific ways. It is the remit of the data governance function to ensure that data is captured, processed, managed, and reported to authorities in compliance with applicable laws, whether those laws relate to privacy (such as GDPR), data security (such as CPS234) or, in this case, the Anti-Money Laundering and Counter-Terrorism Financing Act 2006.

A robust and empowered data governance function within the bank would have given them a better chance of meeting their data management and reporting obligations. In fact, there were multiple opportunities to fix the broken processes that led to this massive fine, but these opportunities were not acted upon.

Later in this book, we will look at how the data governance function can bring correction to poor data processes in a number of ways:

1. reactively: by capturing and then addressing data governance issues as they are uncovered in the course of any information worker's day job;
2. proactively: by deliberately looking for data governance issues that might be reducing the value that the organisation can derive from its data assets, or that are impacting business operations, efficiency or, indeed, its data management obligations;
3. as part of project methodology: by establishing a data governance component within the DNA of every IT project, to maximise opportunities to improve data governance as part of any software upgrade or software development lifecycle.

Looking more closely at what happened at Westpac, the correct alerts for AUSTRAC were designed into the LitePay application that was the platform used for many of the questionable transactions that should have been reported — so the project methodology aspect (aspect 3 above) was correctly observed. The design specifications of the app included the right approach to data governance and, specifically, their data reporting obligations. It is worth noting, however, that the AUSTRAC alerting code wasn't put into production until three months after the app was launched in the marketplace. Apparently, the risk of a $20 million fine per breach was not enough to delay the project launch! There would be a senior manager somewhere in the bank who made the call to take the project live without the AUSTRAC reporting component working, probably in order to meet a deadline and perhaps collect a bonus; that executive's decision actually exposed the bank to a fine potentially in the hundreds of millions of dollars.

However, even after it was put into production, the code for these alerts didn't work. In fact, LitePay operated for over two years before someone asked the obvious question, "Why have there been no alerts generated for suspicious transactions in two whole years?" (This highlights that while data governance was factored into the *design* of the LitePay application, the software *testing* prior to go-live did not scope data governance into their testing cycle — a blind spot in the testing aspect of the project methodology.)

That observation created the reactive opportunity (aspect 1 above) — to fix a broken process or technology, discovered as part of someone's day job that was causing a failure in data governance. This opportunity also was not acted upon, even after the bank received a warning shot in the form of a $700 million fine brought by AUSTRAC against another Australian bank, the Commonwealth....

At the time, it was the biggest fine in Australian history. In a sadly familiar story, the bank failed to manage and correctly report data relating to suspicious transactions. The Financial Review reported that "Since its settlement with AUSTRAC, Commonwealth Bank has been spending around $1 billion each year on improving risk and compliance systems." It appears that this record fine got the attention of senior management and the commensurate investment in better managing its data — but didn't get sufficient attention at Westpac headquarters just down the road.

So, how does an organisation avoid these kinds of disasters arising from failures in data governance?

The Need for Cultural Change

A successful data governance initiative — which means moving from a current low state of data governance maturity to a higher state, and then sustaining it — is essentially a cultural change. In a large organisation such as a bank, insurer, telco or government department, this is no trivial matter. Many organisations have adopted new technology relatively quickly over the last 50 years, but the cultural change to a truly data-mature organisation does not automatically follow.

To illustrate, technology has allowed us to move from typewriter-generated bills (think of the telephone bill your grandparents used to receive in the mail — literally hand-typed on a typewriter until the first electronic printers became available) to printed bills, to emailed bills, to 100% online-only virtual bills and now, to automatic deductions where your bills are paid electronically and you just get an FYI. However, the transition in thinking — from where data is a by-product of a business process to where data drives constant refinement and improvement of the business process — does not automatically follow.

Instead, organisations get comfortable with the entrenched culture, and smart people with good ideas about how to improve processes — or people who see something that's wrong and want to fix it — are often sidelined, ignored, "managed out", or end up so discouraged and disillusioned that they manage themselves out.

The cost of these lost opportunities to improve organisational efficiencies should be regarded as unbearable, even without the massive fines and associated reputational damage that we've discussed already. Indeed, these are just the tip of the iceberg in terms of the cost of poor data governance to Australian businesses.

The Impact of Poor Data Governance

We've seen a number of examples of how damaging bad data governance can be to an organisation, both financially and in terms of reputation. However, a lack of data governance can incur many kinds of commercial and opportunity costs. These include the following:

I once sat in a meeting with a general manager of a very large Australian company. He described how he had a small army of maybe 20 analysts who spent a lot of their time crunching numbers in spreadsheets for his monthly performance reports. All the GMs would come to the monthly management meetings, with their reports from each of their own little army of 20-odd analysts. And then, they would spend the first half of the meeting trying to understand why everyone had different numbers and whose reports were right!

The operational cost of that fiasco was huge — perhaps the equivalent of fifty full-time employees, wasted! (20 people x 10 teams x 25% of their time equals 50 FTEs.)

Most organisations spend enormous amounts of time collecting, correcting, and managing data. We should not underestimate the direct commercial impact that this time cost has on businesses. Thomas Redman outlines the cost of poor data management in a 2016 article for the *Harvard Business Review*:

"$3.1 trillion, IBM's estimate of the yearly cost of poor-quality data, in the US alone."

Given that the US economy in 2016 generated $18.7 trillion, the cost of poor data management equates to 16% of GDP — which is a horrendous drag on any economy. Given that data plays an ever-increasing role in the economy, this drag will only get bigger if the root causes are not addressed. For data-intensive organisations like banks, insurers, telcos, government — in fact, just about anyone outside trades and small business — this hidden cost is likely to be even higher.

So how is this hidden cost incurred? We can see where this cost arises when we consider the steps required to correct data errors using this example from the same article:

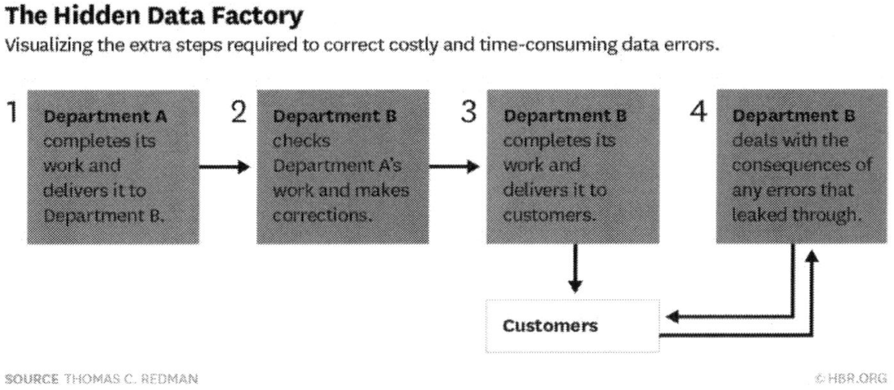

The multiple manual steps and rework result from a lack of trust in the data and poor data quality, both of which are factors in data governance. Take this and multiply it across almost every data exchange activity in the organisation, and you can see what kind of an impact it has — and the size of the opportunity to fix it!

Issue #2 – Having Lots of Data (and No Idea What to Do with It)

Both businesses and individuals are creating volumes of data that are increasing exponentially. At the end of 2020, it's estimated that the total data volume stored around the world added up to 44 zettabytes (a zettabyte is a trillion gigabytes). This will grow to over 250 zettabytes by 2025. More data was generated in the last two years than in the entirety of human history before that. Now, much of this is TikTok and YouTube videos, but much of it is also generated by business activities.

The rise of sensors in everything from your car to your fridge and the advent of the Internet of Things (IoT) or "connected everything" is driving much of this increase in data — but it's not being created or collected for no purpose. Sensors in your car can predict when servicing is required on certain parts and save you a lot of money through preventative maintenance. Then, when the manufacturer collects that data from across the entire fleet that they've sold, they can analyse that data and use the outputs to fix manufacturing issues that used to turn up as warranty claims or in crash statistics. There is a huge benefit in collecting and processing all this data.

But the benefit is not unconditional — you have to know what problem you're trying to solve or what insight you're looking for. And you have to know how you can and can't use the data: some data may have been collected subject to privacy constraints, making it illegal to use for certain purposes. Understanding what you can and can't do with data falls under data governance. And, in summary:

Data that is ungoverned is unusable.

This is because, without knowing the source, quality, and legal and ethical constraints on the data, you can't safely use it.

Organisations with huge volumes of data that are uncataloged may be sitting on a gold mine and have no idea that they are.

Moreover, the sharing of datasets under both free and commercial models means that more data is available to businesses than ever before — from real-time traffic data to customer sentiment to location-based data on businesses and consumers in ever greater detail. This data can be used to uncover insights in behaviour to drive growth, revenue, customer experience, and other potential benefits. Companies that know how to integrate this data with their own data to enhance its potential will outperform those who are restricted to working only with the data they themselves produce.

But without effective data governance in place, your ability to generate value from all this data is severely undermined.

Issue #3 – The Inability to Make Key Decisions

Good business decisions are increasingly dependent on having good quality and reliable data. I've seen many organisations where a lack of trust in the data has paralysed the ability of leadership to make key strategic decisions or has led to costly delays in decision-making that yielded competitive advantage to faster-moving players. This can happen at both the senior management and operational levels.

While it's unrealistic for organisations to achieve 100% data accuracy, access to contextual information about data quality allows a realistic level of confidence in the data available to support a key decision. Knowing — or having some idea of — the accuracy, completeness, and timeliness of the data allows you to determine whether it's fit to support a decision or initiative. Knowing this ahead of time can significantly reduce the cost of poor business decisions and the number of projects resulting in poor outcomes due to unreliable data.

Issue #4 – Lack of Compliance with Regulations

Every year, we're seeing governments on the local, national, and international levels introduce new data regulations. One of the more well-known is the EU's General Data Protection Regulation (GDPR). If you've ever visited a website that asks your permission to use cookies or tells you in detail about how it collects data and how it will be used, you're seeing GDPR in action.

We're seeing a massive wave in regulatory changes like GDPR taking place around the world. Brazil, Australia, and even individual states in the USA like California, are rolling out their own versions of data privacy laws. Failure to comply with these regulations can lead to severe penalties — up to €10,000,000 or 2% of global revenue in the case of GDPR (for a minor offense). Google, British Airways, H&M, and Marriott have all received fines in excess of €10,000,000 for GDPR violations relating to personal data. Marriott was initially fined almost €100,000,000 for a data breach that affected 339 million guests' personal data, of which only 10% was related to European residents. Interestingly, it was found that the vulnerability of personal data came from the systems that Marriott had inherited when they purchased the Starwood Hotels Group in 2014. The EU's Information Commissioner's Office (ICO) found that Marriott had failed in its due diligence to identify data security issues in the Starwood IT systems when it bought the company. The fine was eventually reduced to $23.8 million — still far more expensive than a proper security audit of the IT systems they inherited from Starwood would have been!

The Data Governance Framework is the place to capture and publish these requirements and then integrate them into an overall data management plan that ensures compliance across the organisation. It is critical from a regulatory standpoint to approach this proactively (it's also much cheaper) — rather than allow business as usual and then hope that all will be well.

Issue #5 – The IT Department Becomes a Bottleneck in the Business

There was a time when data was strictly the remit of the IT department. Data was incidental to most roles — a way of doing the same job a little faster and easier. Those days are over: for most organisations, data is integral to every business process and to almost every role in every department. Hence, data is no longer the remit of IT, it's the remit of every worker across the organisation.

Similarly, data governance cannot be thought of as a specialty within IT, or something for IT to fix. When this approach is taken (or is allowed to linger, since it was always done that way), IT ends up responsible for trying to fix the data issues in every part of the organisation — every business process, every action, every report, every decision. It's neither feasible nor economical to try and manage data in this way. It creates a dependency upon IT for every data-related process in the organisation — which is pretty much all processes! IT then becomes a bottleneck for every change, every innovation, and every decision.

The data governance function needs to span the entire organisation just as processes that create and use data span the entire organisation. Later in this book, we will look at a sample organisational structure that supports this approach and ensures that data governance doesn't have to be a bottleneck in your business.

Poor Data Governance Impacts Digital Transformation

Organisations are also beginning to realise the impact that poor data governance is having or will have on their plans for digital uplift through initiatives like automation, AI, and machine learning. A recent study highlighted that almost 90% of information workers pointed at data quality issues as the reason their organisations failed to successfully implement AI and machine learning.

In chapter four, we'll look more closely at the options that are available to organisations with respect to competing in the Fourth Industrial Revolution, with or without their data governance house in order — but for now, let's just say there will be winners and losers. It takes time and additional expense to get the foundations right; but if that investment is made, the resulting "house" is better placed to withstand the oncoming storm of disruption. To build a house on sand is cheaper and faster, but we know how that story ends.

CHAPTER 3

The Benefits of Doing Data Governance Well

A study by the consultancy Forrester in October of 2020 quantified some of the benefits of getting data governance right. Some of the highlights of that investigation include the following insights:

- For one organisation, faster access to data sources and better information about the data (metadata management) drove a greater than 50% increase in the productivity of their business analysts and provided a faster turnaround time for reports and insights. This alone was quantified at $5,256,583 of value over three years.
- Another company in financial services saw an acceleration in information request response times in areas like supply chain and inventory, which drove savings of $7,762,585 over three years.

And this doesn't even take into account the increased productivity of the IT people working behind the scenes to support these processes.

So there are hard, measurable, dollar-quantifiable benefits to getting your data governance right (we'll look at some approaches that can help your organisation put a dollar figure on the potential value in better data governance towards the end of this chapter). There are even greater benefits to be realised when we consider digital transformation — AI, process automation and decision automation. But even without the compelling argument of the potential value in these components of digital transformation, it's also worth considering some of the most immediate benefits of good data governance.

The Benefits of Good Data Governance

The real benefits of data governance come from being able to have *confidence* in your data, from being able to use it *safely* for analytics, insights and decisions, and from better and more accurate *decisions* based on facts rather than gut feel or a finger in the air. The value of data increases exponentially as it is used to drive strategic decisions across the organisation. Also, at the day-to-day level, effective data governance literally drives down the cost of every process in the organisation that uses data — which is pretty much every process in the organisation. Most important of all, good data governance lays the foundation for your organisation to reap the benefits of digital transformation and the next industrial revolution.

Let's examine some of these benefits in more detail.

Benefit #1 – Better Business Decisions

Data that is understandable, of good quality (we'll talk about what "good" or "fit for purpose" means later), and accessible to decision-makers is fundamental to sound business decisions. When your data is untrusted, of questionable quality or veracity, or is difficult to access or difficult to understand, it disempowers rather than empowers sound, fact-based decisions. This will become even more important in the future as AI and machine learning begin to play an ever-increasing role in decision automation.

Benefit #2 – A Data-Driven Culture

Many leaders talk about the need for a data-driven culture. However, if your data is not well-managed, trusted, and available to support business processes and decisions, you're not even at square one. An organisation that has got its data governance house in order can confidently move towards a data-driven culture. Executives know that they need trustworthy data and analysis to support their initiatives and decisions. Good data governance provides the necessary foundation for a data-driven culture.

Benefit #3 – More Confidence in Business Decisions

Data that is well understood, available, and of good quality creates a level of trust in the reliability of the data, which in turn creates a greater sense of confidence in the decisions that you make. Confidence in a business decision drives focus and a sense of common purpose across the organisation. And ultimately, those decisions will be of greater benefit to the company because they're based on accurate reflections and measurements of the real world.

Benefit #4 – Access to Deeper Levels of Insight

Data that is well understood can provide profound insights into everything from how well your business is operating, to predicting the movement of markets and even the "next big thing." As an example, a friend of mine, Elisa Choy, has successfully predicted everything from reality TV show winners to the outcomes of elections, using novel approaches to analysing "big data" and sentiment (you can read more on her website, mavendata.com.au). In the following chapter, we'll look at how "big data" is driving new frontiers in understanding patterns in nature and in human behaviour.

Companies that have their data governance house in order will be able to gain new and deeper insights into customers, market trends and other key enablers of their business success, much more powerfully and reliably than will organisations that are still struggling with a business case for an investment in data governance!

Benefit #5 – Reducing Data Risk

Working with data always carries a degree of risk. These risks can be broken down into a number of areas: the risk of data being misinterpreted, the risk of bad data or poor-quality data steering you to the wrong conclusion, the risk of someone in the organisation accessing restricted data such as sensitive personnel files or customer information, or the risk of a data security breach leading to the kind of incidents we have already discussed, including reputational risk for the organisation.

Good data governance significantly reduces all of these risks. Indeed, one of the key goals of data governance is to reduce data risk. We will look at data risk and managing data risk in great detail in chapter 9.

Benefit #6 – Reducing Operational Risk

In some industries such as financial services, well-managed financial data is imperative for managing the risks that are involved in doing business in their markets: underwriting credit, setting exchange rates, predicting future interest rate moves and currency moves, and ensuring compliance with capital requirements. As businesses become more and more digital, the kinds of insights into risk profiling that these financial services firms have mastered will become relevant to more and more business models. The more reliable and accurate the data, the more reliable and accurate the risk models based on that data will be.

However, this does not only apply to banks and insurers. Any business with an e-commerce site is continuously collecting customer behaviour data on their website, such as which items are being looked at, which items are being purchased — or looked at and not purchased — and where in the sales funnel people are dropping off. Changes in any of these behaviours, if detected, can surface insight into a change that is emerging as a threat to your business success.

Benefit #7 – Operational Cost Savings

Recall the situation I mentioned earlier where about 50 FTE equivalents were wasting their time crunching data for the monthly GM meetings. We also looked at the staggering cost that these kinds of practices inflict upon the economy more broadly. But I'll call out one of the key findings again: inefficiencies due to poor data management, poor data knowledge management, and poor data governance can add up to 15% to an organisation's operating costs. Imagine the impact to your bottom line from eliminating a cost like that!

Benefit #8 – Regulatory Compliance

We've already seen a number of examples of very expensive breaches due to poor data culture and data governance. But there's another impact even if your company is not subjected to any fines. Poor data governance, if suspected by a regulatory authority, can place your company on the radar of regulators and invite further scrutiny. This additional scrutiny comes with a higher operational cost as you need to spend additional effort ensuring 100% compliance, with a huge reporting overhead, to avoid a penalty and give the regulatory authority confidence that your house is in order.

Benefit #9 – Preventing and Mitigating Data Security Breaches

When we hear the term "data breach", we often think of the most high-profile cases. The Sony PlayStation data breach is an example: In 2011, the company's online gaming network experienced a breach that led to the theft of data from 77 million user accounts. This prompted the company to shut down a profitable online network, leading to financial losses. Users could no longer purchase games or access online services that they had paid for.

Once again, we saw the reputational impact of a data breach. Unhappy customers flooded message boards and forums to complain about the issue. Sony eventually had to give away free products in an effort to stem the outrage. Fortunately for the company, the outrage did subside, and they're still one of the dominant players in the videogame industry. But this example shows how poor data security can lead to challenges for a company.

Your data governance policies will define different levels of sensitivity for different data sets and based on this, will also define which roles and which people should have access to each dataset — both internal and external parties. These access permissions are enforced by software, but they have to be defined first.

Data governance also includes having clear procedures in place to respond to a data breach. In 2019, The Sydney Morning Herald reported on a case of a much smaller scale. An employee of a real estate company made the simple mistake of CC-ing 300 customers in an email when they were supposed to use the BCC field instead. The result was that 300 customer email addresses landed in the lap of every recipient of the email — a breach of legislation protecting personal information.

One of the customers notified the company, which launched an investigation. Thankfully, no damage was done through the nefarious use of those 300 email addresses. However, the incident forced this small business to spend weeks — and over six figures — working to undo the damage, as they had no process in place to deal with such a breach. The high cost was incurred because they had to hire consultants to create policies on the fly, while simultaneously trying to follow those policies to the letter to ensure maximum compliance with the legislation, and while also undertaking the damage control specified by the policies as the policies for damage control were being created! With a data breach process in place, it would have been a much simpler case of "following the protocol", and the cost would have been much less.

Benefit #10 – Data Governance Allows for Easier Sharing of Insights

The corollary of data access security is data sharing. The ability to easily access data from every business process within (and even outside of) an organisation can rapidly drive new insights for curious analysts and data scientists with a mandate to increase efficiency, drive automation, and discover new trends, correlations, and insights.

For example, your web design team may collect data (from the web logs in IT security) to discover the most successful customer journeys towards a web purchase and then optimise the user experience based on these insights.

However, legislation relating to data privacy (such as GDPR) means that you have to be careful who has access to that data and who should be able to access summarised or masked data but not see detailed records. In the example above, people outside of IT security should not have access to personally identifiable data (such as the IP address of the purchaser).

Your data governance initiative defines the principles that are used to decide which datasets can be shared with which parts of the organisation for what purposes. You can't anticipate and determine every possible request, but with the data security and data sharing principles clearly defined, every data access request doesn't have to trigger a slow and complex workflow to approve or deny access (or to approve access subject to certain constraints).

The Business Case: Putting a Hard Dollar Value on Data Governance

Given all of the benefits that we have looked at above, you would think it wouldn't be too difficult to actually put a hard dollar value on those benefits. However, many advocates of data governance struggle to articulate its value and struggle even more to put a hard economic value on it. We can, however, analyse some of these benefits in terms of actual time saved and actual cost avoidance, which can help support a business case for a data governance initiative.

To establish a business case for data governance, you need to establish the current reality of time wasted and opportunities lost as a result of not having the right data governance practices in place. To do this, you need to get a bit into the details of the life of typical information workers in your organisation.

You need to select a representative group across the organisation that includes business analysts, the people responsible for reporting, as well as data-intensive roles from every business function (or as many as possible) such as marketing, finance, sales, and every significant function within the organisation. To help with this, get ahold of the organisational chart and look at every area where data is used either for reporting (such as financial reports), decision support (such as risk, pricing, etc.), operations, and so on. You will also need to have representatives from IT: people who work with and look after data warehouses, data lakes, BI tools, ETL tools, dashboards, and so on. Include "innovation hub" style teams that work with data and any data science, AI/ML, predictive analytics, and process automation teams. Because organisations across different industries are so diverse, it's hard to put an exhaustive listing in a book like this. But your org chart will provide the overview of what the organisation looks like and where data is being used. By asking the right questions to a few people across the organisation, you'll be able to figure out the best representative sample of information workers who will form the group that you will use to measure the potential value in data governance for your organisation.

Once you've established who is in this group, use the following questions to put a quantity of time wasted or opportunities lost for each category of question. You can then extrapolate those figures across the entirety of the organisation to arrive at a reasonable estimate of the return on investment of uplifting data governance across the organisation — a hard benefit that you can use to support a business case.

Questions to Help Support Your Business Case

Not all of these questions will be relevant to all of the people in your representative sample, but most of them will be relevant to many. Ideally, ask these questions in a 1:1 conversation or small group workshop (whether in the real world or online), rather than by email. This helps you gauge the level of intensity of frustration with the current situation and can provide some very powerful anecdotes and quotes that you can use in your business case.

Feel free to refine and adapt these questions as you learn more about what matters and where the key shortcomings are. The questions are:

- How much time is spent trying to find the correct data source to answer a business question or query?
- How much time is spent checking to see if a report already exists to answer a question you have been asked to solve?
- How much time is spent creating a report or running an analysis and later you find that it had already been done by someone else, or something very similar existed that you could have used to save time?
- How much time is spent trying to understand the source of a report or dataset, or the logic (business rules) that were used to transform that data, to see if it is appropriate for a given purpose?
- How much time is spent checking on the data quality of any data that you intend to use?
- How much time is spent "double-checking" data or reports against other data because one data source is untrusted or the quality is unknown?
- How much time is spent trying to understand the definition of a data element or report item, tracking down someone who knows or having to trace the data "upstream" to find out where it came from and how it has been transformed to get to where it is?

- How much time is spent looking for the owner or subject matter expert for a dataset or data source, to clarify a question about the data?
- How much time is spent trying to figure out if some privacy, legal or ethical limitations apply to a dataset that someone wants to use?
- How much time is spent trying to find the right person to answer any of the above questions or waiting for an answer for any of these questions?
- For data engineers and ETL developers: How much time do you spend answering questions from people about data sources, what the data means, where it comes from, is it up-to-date, are there data quality issues, and so on?
- Do you know of examples where poor decisions were made because of insufficient data, lack of timely data, or poor-quality data? Is there a way to estimate a dollar value for the impact of the poor decision (e.g. poor targeting of an offer that meant wasted marketing spend on the wrong audience)?
- How much time is spent checking figures in spreadsheets and reports because there is a question or lack of confidence in the numbers?
- Are there business decisions that are delayed because we're waiting for the right data or waiting for verification that the data is OK to use? Is there a way to put an estimated cost on the delay (e.g. value of the decision, say $1M of benefit across 2 years, multiplied by the delay, say 2 months, would give 2/24 x $1M = $83K. Calculate or extrapolate this to the number of projects that are delayed waiting for data, waiting for data to be checked, or due to lack of confidence in the data)?
- Are there teams that actually duplicate the same kind of work because one team trusts one data source while another team prefers a different data source to answer essentially the same question?
- Are there projects that have been delayed or have missed certain deadlines because the data was unavailable or of unknown quality or not fit for the purpose? What was the financial impact on the project, and what was the impact to the business of the delay to the go-live date?

Also, consider organisational risk arising from data practices. You may not be able to put a dollar value on this, but it warrants the attention of senior management regardless.

You're asking: Is the organisation exposed to any risks as a result of poor data governance? This could include:

- risk of reporting incorrectly or late to regulators
- risk of errors in estimates of key business metrics or financial ratios
- risk of sensitive data such as PII being exposed either directly or through partner organisations that you have to share data with
- risks due to the absence of processes to manage these risks

All of these questions highlight areas where good data governance will provide massive benefit: in time saved, in errors avoided, and in all of the ten areas of benefits I've highlighted above in this chapter.

For each question, multiply the time lost by the estimated "total cost of employment" (TCE) for the person whose time is consumed (often, you can get a standard figure from HR, such as "average TCE is assumed to be $100,000 p.a.") — then multiply this by the estimated number of people affected in this way.

Document the methodology you have used to arrive at the total impact of poor data governance, as you will need to socialise this among senior management (people with budgetary discretion to support an initiative). Ideally, put the figures in a slide deck along with some powerful quotes and anecdotes from people you interviewed that highlight how poor data quality and governance impacts their role, their efficiency, and their productivity. Add a slide or two highlighting any risks identified and, if possible, validate this section with someone who specialises in Risk in your organisation — they may be able to add their own observations as well.

Work with an advocate in senior management to bring your findings to the attention of the relevant people to get buy-in for an initiative to address the financial and risk impacts of poor data governance in your organisation.

Data Governance and Data Exchange

We've looked at some of the key benefits that good data governance creates at the company level, but there are also benefits far beyond the operations of just your own organisation. Data sharing and intercompany data exchange can unlock additional value and, in response to this, we're seeing the emergence of data "marketplaces" like Data Republic in Australia. We're also seeing "open data" initiatives both in the government and private sectors. We're beginning to see research data from universities shared with colleagues on the other side of the world to boost research and reduce the cost of replicating an experiment.

All of these initiatives support the creation of value from data and data sharing in new and exciting ways. In fact, in 2015, we crossed a very interesting tipping point: in that year, the exchange of data added more value to the global economy (in terms of global GDP) than the entire global exchange of physical goods! A McKinsey study from May 2019 notes that in just two years (from 2017 to 2019) the percentage of companies forming data-related partnerships rose from 21% to 40%. Even business competitors are sharing and connecting data, with the percentage of companies sharing data with competitors rising from 7% to 17%.

In fact, McKinsey estimates that connecting data across institutional and geographic boundaries is already adding roughly $3 trillion annually in economic value across the globe. But the value that can be created through data exchange is critically dependent on the quality of data governance.

CHAPTER 4

Data Governance and the Fourth Industrial Revolution

You may have heard the phrase "data is the new oil". Just as companies made fortunes off the processing and selling of oil in the past, companies today are making fortunes off the processing and monetisation of data.

Indeed, many of the most valuable companies in the world have only one input and output: data. Google and Facebook are the most obvious examples. They are fundamentally data companies, or started out that way. For many years, the only product they had was data and algorithms which leveraged that data to provide a service that people wanted (search results, and friend and content recommendations). Now, they make huge amounts of money exchanging the data that they collect with other companies. They also leverage their data to enable companies to advertise more effectively not just on their platforms but across the digital marketing ecosystem. For these modern tech giants, data is big business — but getting value from data is a proposition that's relevant to companies of all sizes.

To put it another way, data is the raw material that is going to drive the next economic revolution. It's through the effective use of trustworthy data that we can enhance productivity and create more efficient organisations. So, much like the discovery of oil, the new data-driven forms of business will revolutionise the global economy. However, there are some very interesting differences between data and oil...

First, data is far more widely available and is becoming cheaper and more accessible. Whereas oil has been always a finite resource, the amount of data available in the world has a limitless ceiling for growth.

Second, to move oil from one part of the world to another is a slow and expensive exercise, requiring huge tankers specially designed for the task. On the other hand, data can be sent from one side of the world to the other at literally the speed of light and at almost zero cost.

Third, you can't take oil and magically copy it to make more oil. But data can be replicated infinitely for very little cost.

So, data actually has even more potential than oil to exponentially transform the way we work and live. That is why the World Economic Forum and other observers are calling it the Fourth Industrial Revolution. This revolution will go on for far longer than any we've seen before and lay the foundation for further revolutions such as AI and process automation — all of which will increase the efficiency of human labour (time) even more than the first three industrial revolutions did!

The Four Productivity Revolutions

To help understand the impact that will result from what has been termed the Fourth Industrial Revolution, let's take a journey back through time and consider the first three productivity revolutions — the key advancements in human history that have accelerated living standards to levels unthinkable in previous generations.

We'll start by looking at the First Industrial Revolution, which was driven by an incredibly disruptive new technology: the steam engine. To help understand what happened, we'll look at one industry that was famously impacted: cloth weaving.

Once upon a time, making fabric for clothes involved the use of a complicated loom machine driven by human energy — much like the one this young man is using:

FOOT-POWER LOOM WEAVING

Fig. 32—Threading the Heddles

Using this machine was an entirely manual task, with the operator needing a certain set of skills as well as sheer muscle work to create the desired product. And this was reflected in the price of the cloth: to purchase a suit back in the day when this young man was working would cost the equivalent of 18 months of a typical worker's wage. This is because it took literally that many person-hours to produce the final product (plus some margin at each point in the supply chain).

However, the introduction of steam power changed that entire industry. By connecting the loom to automated power rather than foot power, the muscle work part of the loom could be automated. On top of this came another invention by the Frenchman Joseph Marie Jacquard: the Jacquard loom. It provided a method for changing the cloth colours using wooden boards with a pattern of holes, allowing the pins to move up and down as the cloth was woven. In this way, the machine could create the pattern that used to require our young friend's manual switching of cloth colours. (In fact, you could argue that Jacquard's loom was the first use of punch cards, a method we were still programming computers with almost 200 years later!)

So, what happened? Our young friend's entire industry has been disrupted, and he's out of work — made redundant by technology (doesn't that sound familiar?). In fact, the Luddite movement emerged at this time as a response to this disruption. This group became famous for attacking factories and smashing automated looms in an act of protest against the disruption of their industry and the loss of their jobs.

However, two other things also happened as a result of this disruption. Firstly, the price of cloth began to drop dramatically as fewer people needed to be paid to produce the cloth, and so, good quality clothes became cheaper and more accessible. This productivity gain can be extrapolated to today when a suit can be purchased for about a single day's wage rather than 18 months' worth!

Secondly, the young man who was made redundant is actually freed up from boring and repetitive work to undertake something more creative and more value-adding to the overall economy. Given his skill with cloth and with his hands, he could go on to become a tailor, a shoemaker or perhaps a blacksmith — jobs that, in his day, would not only pay more but would also be more fulfilling as they were more demanding of specific skills. And the value that he can offer in one of these jobs would have been higher than the value he was creating spending days weaving cloth on a manual loom. In fact, a study in 1983 by Peter Lindert and Jeffrey Williamson calculated that real wages, or purchasing power, doubled for English blue-collar workers in the 32 years from 1819 to 1851. Even the lowest socioeconomic strata saw their real incomes rise by 70% in the same time frame.

So, the First Industrial Revolution saw massive disruption and a shake-up that impacted many industries and jobs. However, it also freed many people from boring, manual, repetitive work, freeing up their time and skills (note that re-training in new skills might be needed). This allowed them to take on more creative and interesting work, which also delivered more value to the overall economy. It's a pattern that we will see repeated in all the productivity revolutions, including the current one!

The Second Industrial Revolution is regarded as being the invention of electronics and the introduction of computerisation into the workplace. Perhaps my favourite example of the effect of this revolution comes from the picture below:

Once upon a time, highly intelligent and capable women (yes, it was a very sexist workplace) really did spend their days sitting in front of a patch panel and moving quarter-inch plugs around to complete phone calls. This actually was an enormous time and cost burden on businesses because of an inefficient process — a cost that of course was ultimately borne by the customer, as the price of whatever good or service was produced had to include the cost of this human labour. We also see a huge waste of human talent in the sense that these women could provide so much more value to the economy in other capacities.

Electronics and computerisation allowed for a new level of automation that handled tasks like connecting phone calls automatically. Once again, the workplace was disrupted by technology: many roles were made redundant by electronics, with massive job losses as a result. However, these job losses actually meant that these talented and intelligent young people were freed up to fulfil higher aspirations and add far more value to the economy. They could spend their careers as engineers, lawyers, teachers, doctors, or indeed, prime ministers!

The Third Industrial Revolution comes on the back of something that dominates all of our lives today — the internet.

I remember my parents forking out a very large sum of money to purchase the entire set of World Book Encyclopedias from a door-to-door salesman to benefit their kids' education. Today, we can access all of the information that we will ever need from a computer or even a smartphone, and most of it for free! The internet opened up the world and made it easier for people (and businesses) to communicate and collaborate. This was quickly followed by e-commerce as online marketplaces disrupted existing retail models and supply chains, bypassing the traditional bricks-and-mortar model. We saw companies like Blockbuster and Borders booksellers go under as their business model was made redundant. We also saw massive new industries and job types emerge which had not been imagined only 20 years ago — search engine optimisation, digital marketing, and so on.

But what comes next?

The next productivity evolution will be a data-driven one. And we're already seeing the beginnings of this revolution. The introduction of technologies like artificial intelligence (AI) and machine learning (ML) all rely on data to operate effectively. With data, we're going to see a new wave of process automation that will go beyond anything that we've seen before. And much like the revolutions that came before, the rise of this new kind of automation — decision and process automation — will result in massive industry disruption and, in many cases, the redundancy of old jobs.

The next productivity revolution is upon us — and its raw fuel is data. This is why data governance is perhaps more important now than it has been at any point in our history.

How This Productivity Revolution Actually Works

Way back in 2008, before the phrases "big data" or "data is the new oil" had begun to circulate, *WIRED* Magazine ran a cover story titled "The End of Theory: The Data Deluge Makes the Scientific Method Obsolete". If you're interested, you can read the story here — https://www.wired.com/2008/06/pb-theory/.

This article was seminal in defining and explaining how ever-increasing volumes of data could lead to massive gains in productivity. It's instructive to unpack this fascinating article.

The article first explored the concept of the scientific method. In the pre-Petabyte Age world, the scientific method operates as follows (in very simplified terms):

1. Start with observations.
2. Create a hypothesis that explains the observations.
3. Test the hypothesis by making more observations.
4. Use those further observations to validate (or invalidate) the hypothesis.

The article highlighted how this had always been a painstaking process. It takes time to make observations, craft hypotheses, and test them. And much of this time came from the need to collect data.

For example, data points — observations — might be gathered over a number of experiments and collated in some format, such as this chart:

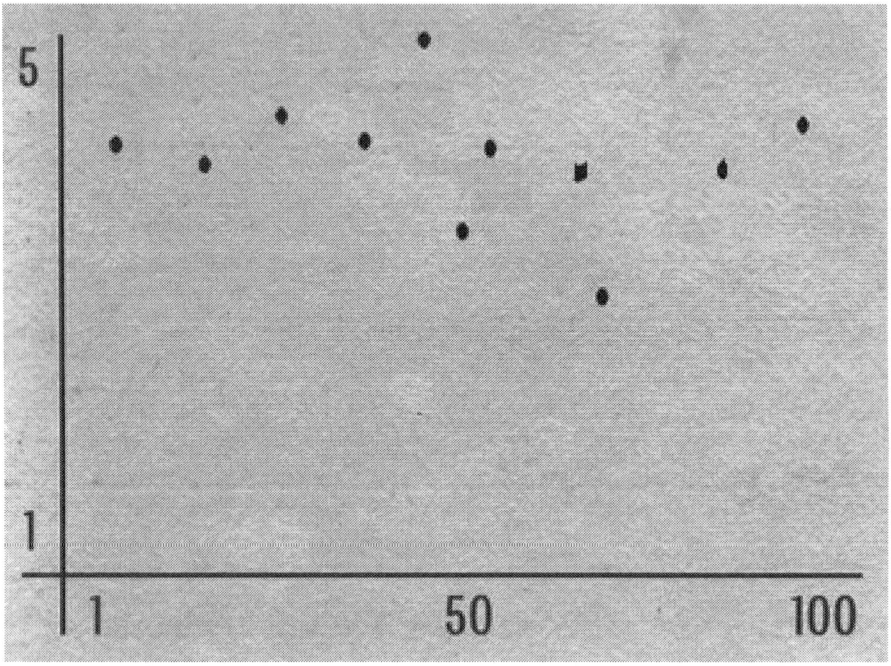

The dots you see are the data points that we have available. From those dots, we could hypothesise a slow, linear upward trend, or it could indicate something more like a sine wave with a flat or slowly falling trend. But we don't have enough data to draw a trustworthy or provable conclusion. We would need to create more observations — more data points — which takes both time and cost. To use a real-world analogy, these observation points might be measurements taken inside some part of a jet engine, after every landing. We can't observe the jet engine in flight, we can only take measurements when it's on the ground. And from that, we can try to hypothesise what is happening with the engine in flight.

The advent of Big Data means that so many more observation points are available. These come from sensors, mobile devices generating location data, and so on. And once the sensors are built and installed, the data is virtually free. This reality, according to the *WIRED* article, has flipped the scientific process on its head by making data more accessible to researchers, analysts, data scientists and others who need it. To return to our jet engine story, the manufacturers are now loading their engines with sensors in key places that collect data while the engine is in flight. Some of these sensors record 5,000 measurements per second, producing terabytes of data from every single flight.

What you get is a dataset, more like the one below, which allows you to massively accelerate the scientific process. Instead of needing to create more observation points through expensive experimentation (such as setting up the jet engine in a wind tunnel), the number of data points available means that the trend simply jumps out at you! Indeed the concept of data science only exists because datasets are now large enough (with enough observation points) that the underlying trends — and therefore predictions — can be made with little more than good (though sometimes very complex) mathematics.

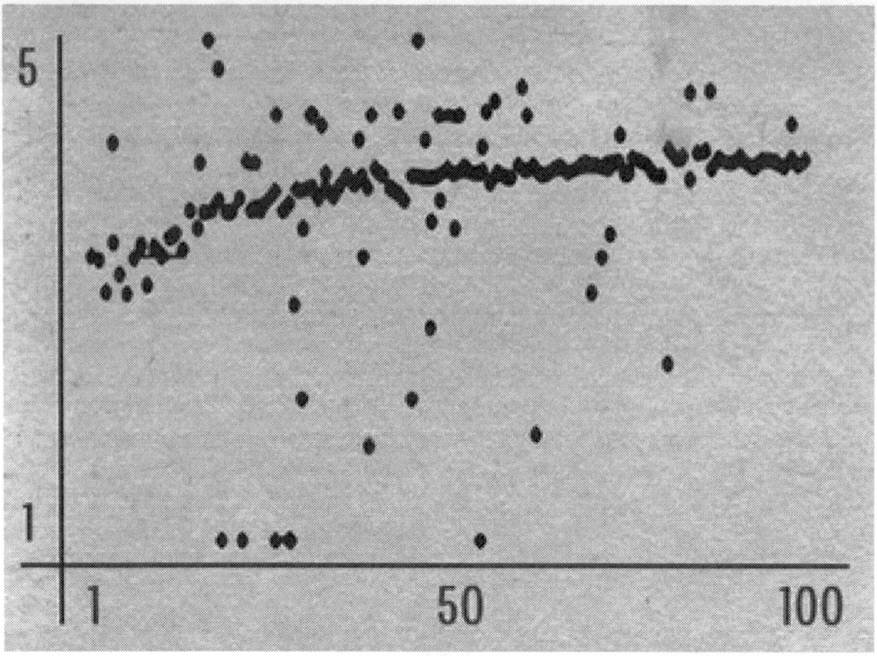

The ability to collect such huge numbers of observations so cheaply is not only disrupting traditional business models but also leading to the emergence of entirely new businesses and business models.

Take car insurance as an example. In the traditional model, the insurer collects a small number of data points, such as the age of the youngest driver, the make and model of the car, where and how it is garaged, and some driving history. On the basis of this data, a team of very smart (and expensive) people — usually qualified actuaries — attempts to price the risk of any policy for every driver and car based on past data for similar drivers, cars, and postcodes. Indeed, one insurer I worked with had a team of over 20 PhD qualified actuaries working full-time on refining and updating these risk and pricing models for their car insurance.

However, new entrants are posing significant challenges to this market today. Rather than collect data about the car and driver, they ask you to install a single sensor in your vehicle. This sensor transmits back to the insurer second-by-second observations on speed, location (and therefore the applicable speed limit), acceleration, braking, and so on. With nothing more than this data, they can price the risk of a policy with a single algorithm. It doesn't matter if you've got an 18-year-old driving a V8 muscle car: if he drives carefully, he's a low risk and is priced accordingly. This approach is not only fairer but also frees up our 20 PhD qualified actuaries to find better ways to use their talents!

Another example of accelerated time to value from better use of data comes from the search for a vaccine for COVID-19. At the time of writing, the world is still in the grip of the pandemic. But we've already seen a team in the UK use a supercomputer and known data to narrow the field from hundreds of possible options down to just a few in order to select a small number of vaccine candidates with a higher probability of success. They effectively eliminated years of research and testing that the old scientific model would require. And they replaced it with a few days of high-intensity data processing.

Where Does Data Governance Fit into This Revolution?

Data will be the source of the new wave of innovation that will affect all industries the world over. It will be the key asset that's responsible for a productivity uplift, the likes of which we haven't seen since the introduction of the steam engine or the internet.

But that, in itself, creates a challenge.

If we fail to manage and govern our data well, we cannot ride the wave of this productivity revolution. We will be working from inaccurate data or data that we don't trust, which means our data will not provide the value that it has the potential to provide — or worse, we could automate and accelerate bad decisions and outcomes based on bad data.

And so, the role of data governance in the upcoming revolution is to create trust and assurance — and safety rails — to enable us to maximise the value that we can get from our data. Data governance is already foundational to gaining trustworthy actionable insights from analytics. In the fast-approaching world of machine learning and artificial intelligence, data becomes even more valuable and so, data governance becomes even more critical.

Indeed, I believe we will see organisations head in one of three possible directions as we see this revolution unfold.

In the first category will be companies who are not anticipating the rapid change to the competitive landscape that automation, machine learning and artificial intelligence will foist upon their industry. They will struggle to come to terms with the new realities as their competitors take market share, whether through better products, better understanding their customers, or simply by offering a cheaper product as they pass on the savings from process efficiencies. These organisations will likely go the way of Blockbuster, Borders, Kodak, and other brands that we have seen disappear as a result of disruption they did not respond well to. Even those that survive will struggle to keep pace with their competitors as the competitors learn to tweak and exploit AI/ML more and more.

In the second category will be those companies that have laid the foundation well by investing appropriately in data governance — ensuring that critical data elements throughout their organisation are documented, understood, and curated — and that this knowledge is available to information workers who need to know how to use the data correctly. They will have processes for ensuring data quality is known, and that a new initiative or project can very quickly ascertain whether a given set of data is fit for the purpose for which they intend to exploit it. As they automate more and more processes through artificial intelligence and machine learning, they will be able to confidently implement their models into production, as the hard work of getting the data right has been done. The value they derive from the crafts of the Fourth Industrial Revolution will increase exponentially, as they employ into future projects the learnings from previous ones — while their competitors are forever trying to catch up.

In the third category will be companies that see and embrace the AI/ML revolution and start to deploy machine learning models, process automation, data-driven decision-making, and all the trends and trappings of the revolution. However, they will find very quickly, after a few disastrous implementations, that their data is not up to scratch. The quality is unknown, its fitness for purpose was not verified, and they will have to go back to the basics of data governance and lay the right foundations before they can properly implement data-driven decision-making, process automation, machine learning models, and so on. They will scramble to get their data governance house in order while their competitors, with a mature data governance culture, leave them behind.

With this in mind, it’s time to dig into data governance in more detail.

CHAPTER 5

What Is Data Governance?

So far, we've looked at the critical importance of data governance to the health of our organisations and economy. We've looked at some tragic examples of the cost of poor data governance and some hard dollar benefits of doing data governance well. We've also looked at how data governance is a crucial foundation for any organisation to be able to gain from the next productivity revolution, and what will happen to companies who don't have their data house in order.

It's time to drill down into the "what" of data governance now that we've covered the "why".

Towards a Definition of Data Governance

A quick internet search will return any number of definitions for data governance. Some are more helpful than others; but rather than try and arrive at a single definition, it's worth taking note of a few.

The authoritative DMBOK (Data Management Body of Knowledge) defines data governance as *"the exercise of authority and control (planning, monitoring, and enforcement) over the management of data assets"*.

Datagovernance.com unpacks this a little further:

"Data Governance is a system of decision rights and accountabilities for information-related processes, executed according to agreed-upon models which describe who can take what actions with what information, and when, under what circumstances, using what methods."

I prefer my own definition, which I have also "sanity-checked" with my own network and people in several organisations I've worked with:

"Data Governance refers to the ***collation*** *and* ***management*** *of* ***knowledge of key aspects*** *of an organisation's data assets, and the management and improvement of* ***processes*** *that determine the* ***quality, usability and***

availability *of those data assets, so that* ***dependent processes and stakeholders*** *are able to* ***safely*** *access* ***fit for purpose data and information*** *in support of ongoing business value creation."*

You can see that I tend to place less emphasis on enforcement and accountability, and more on empowerment through access to information about data assets. (Of course both are necessary, but I find people tend to not like being told what to do, so I focus on making it easier to work the right way rather than telling people not to work the wrong way.)

It's also worth mentioning a much simpler definition from David Dadoun who is another authority on the subject: *"Data Governance is... just preventing people from doing stupid things with data!"*

Another way of simplifying this is to break data governance down into two key components:

1. Knowledge Management
2. Process Management

Knowledge Management

Knowledge management means having (and making available to information workers) key information about the data they can access and will work with. It involves collecting, curating, publishing, and updating the information about data assets that is needed to understand how to use those data assets correctly, safely, legally, and ethically. It's knowledge about that data that empowers, and provides the safety rails for, people who will derive meaning and value from it. If we don't have that context around the data, any conclusions we draw could be nonsensical or even dangerous.

When we look at the Data Governance Framework in later chapters, the sections of Metadata Management, Data Lineage and Data Usability relate directly to knowledge management. Knowing the quality of the data is also important in being able to determine whether and how to use a data asset, so there is a knowledge management aspect to data quality also.

The easiest way of collecting, curating, publishing, and updating this knowledge is with a dedicated "knowledge management" platform, such as a wiki; or a larger organisation might use specialised software called a Data Catalog. (We will look at this in more detail in the chapter on Technology.)

Process Management

A significant part of getting data governance right is establishing and refining the processes that are integral to managing data well. We will look at this in some detail in the Data Governance Framework. But to provide an overview, implementing data governance will include setting up processes for the improvement of:

Data Usability: processes to continuously capture better and better information (knowledge) about data assets and make these available to the community.

Data Availability: processes by which people can easily request access to data and have that request reviewed and approved (or, if appropriate, modified, so that access is given to a dataset that meets their requirements even if this differs from the original request)

Data Quality: I like to highlight three different kinds of processes related to managing data quality in the context of data governance:

1. the reactive aspect: processes to capture and then respond to data quality issues as they are identified in the course of any information worker's day job
2. the proactive aspect: processes that proactively look for data quality issues that might be reducing the value that the organisation can derive from a dataset or that are impacting business operations or efficiency
3. the project methodology aspect: processes that establish a data quality component within the DNA of any project, so that opportunities to improve data quality (e.g. as the data is captured in a new business process) as part of any software upgrade or software development life cycles are identified and maximised

You can take the same three approaches to identify opportunities to improve data governance overall, as well as data quality specifically, as we noted in an earlier chapter.

Data Security: processes by which data security issues can be raised, evaluated, and acted upon.

Architects and Builders

When it comes to how data governance actually works in an organisation, it can be helpful to use the analogy of architects and builders.

An architect creates the overall design of a building, taking into account how it will work and how people use it, and the relevant legislation and design requirements. This includes everything from materials engineering to climate and shadows to ergonomics to physical privacy. They are also influenced by the culture they are designing for, as cultural expectations also affect the design of buildings (take traditional Asian building styles, for example).

The builders' role (in very simple terms) is to do the work of putting things together according to the design, principles and guidelines provided by the architects.

In the same way, with data governance, there will be a team of people who are responsible for the principles, guidelines, and overall design of how data is used, maintained and managed, taking into account best practices, legislative requirements, and the organisational culture. These correspond to the architects, and we'll unpack the kind of roles and role descriptions involved when we look at the organisational structure for data governance.

Everyone who works with data (which is pretty much everyone in the organisation) corresponds to the builder: whatever they do with data in their day-to-day jobs needs to align with and conform to the principles and guidelines provided by the data governance function.

As an organisation becomes more data literate and more mature in its understanding of the value of data and of good data hygiene, it will become as unacceptable to take shortcuts with data governance as it would be for a builder to take shortcuts for convenience, ignoring the stipulations and standards laid down by the architect and the building regulators.

SECTION 2

Understanding the Data in Data Governance

CHAPTER 6

How Data Works in the Organisational Environment

A helpful way to understand how data is created and managed in an organisation is the BIAT model. It breaks down the layers in which data is generated, used, captured, and stored.

- **Business Process Layer** – This is the ***process or activity*** that generates the data as an organisation undertakes its business activities, including how data is used and modified as part of a business process.
- **Information Layer** – This relates to the ***data itself***, which was created, captured, or modified in the business process layer.
- **Application Layer** – In this layer, you have the ***software and applications*** that capture, manage, and store the data. It includes user interfaces, web forms, apps, and all the code that manages the information, supporting the business process layer.
- **Technology Layer** – This is the ***underlying infrastructure*** for the application layer. It includes things like your servers, databases, hardware, operating systems, security measures, and cloud infrastructure.

In this chapter, we'll dive deeper into what each layer of the BIAT model means. We'll also discuss the different types of data that your organisation may collect and manage.

The Business Process Layer

As mentioned, this is the process or activity that captures or creates data or modifies previously captured data. For example, a customer logging into a website for the first time is a business process; a returning customer logging into the website is a slightly different business process. In the first case, a customer needs to provide information, such as their name and email address, and set up a password. This data is captured and stored (in the lower layers) and is then used to support other business processes. A returning customer may just need to provide a login and password to gain access to the website again.

The same process can be seen with, say, a customer phoning a call centre to apply for a credit card or update their address. This layer covers all of the processes that create or modify the data that your organisation will use in one way or another.

Often, one business process will access the information layer (i.e. data captured from previous interactions in the business process layer) and trigger further actions. For example, when you log in to, say, Amazon.com, as soon as you provide your details, the underlying applications trawl through your purchase history to make some recommendations as to what you might like to purchase next (this is called the "next best offer", and a lot of data science and predictive analytics goes on under the covers to bring you the most compelling personalised offer).

That's the business process layer in action for an e-commerce store. Of course, the exact processes will differ from business to business.

The Information Layer

As noted above, all business processes will create, capture or update information. Sometimes, that information is never stored, e.g. when someone gives you a verbal quote or price. But in the age of technology, huge amounts of data are captured from almost every business process. And in the Internet of Things (IoT), machines or sensors generate data that is then passed to other machines for processing. All of the data that is "touched" or created by a business process (whether automated as in IoT or manual as in filling in a web form) is captured and becomes part of the information layer.

For example, data collected from a web form may be captured in the web application as a Java object or some other construct. This is now data in context that, therefore, has meaning and can be used as information — which other processes can then act on.

Another example is data collected in a call centre. Calls are often recorded as mp3 audio files. These can be converted into text, which is then stored in a database. Using text analytics, useful information such as sentiment analysis can be derived from this data.

The Application Layer

The application layer is essentially the software mechanics that manage the information that's created or modified by the business process. It includes the software itself and the user interface or web pages generated by the software. The application layer captures and manages the information that is created or modified by the business process and presents it in a useful way to the people who need to work with it.

For example, a business may use Salesforce as their Customer Relationship Management (CRM) application. This software supports a complex range of business processes, some of which the customer themselves might participate in (by logging into a customer-facing web page) and some of which are internal to the company. The application provides a certain view of selected information to, say, a call centre worker and a different view of information to a sales manager. When designing the software, first, the business processes are mapped and analysed, then the information generated or updated by each process is defined, and then the software is designed to support that business process and to capture, manage and present the information.

The Technology Layer

This layer covers the infrastructure that makes it possible for the application layer to function reliably to gather and use data; that includes servers, databases, storage, cloud-hosted infrastructure, and even things like network connectivity. It also includes middleware and messaging (like APIs) that provide data exchange within and outside of the organisation.

The Difference Between Data and Information

There is in fact a difference between data and information.

The Australian Prudential Regulatory Authority (APRA) defines data as follows:

"Data refers to the representation of facts, figures and ideas. It is often viewed as the lowest level of abstraction from which information and knowledge are derived."

Think of data as a raw fact or observation that has no context to help determine what this fact actually means. As an example, "the temperature was 27 degrees". For data to become meaningful, it has to be interpreted in a context. Using the same example, by adding context, we get: "The maximum temperature in North Sydney on Friday, March 1st, was 27 degrees centigrade." The context allows for the derivation of meaning from a bunch of facts.

And so, information comes from data. Indeed, we can't have information without data, and we can't use data until we turn it into information. For data to be turned into information, that meaning and that context are provided by the data governance function. So you could argue that data governance is what allows you to safely turn data into information. That's not a definition of data governance, it's an effect of data governance, but it's a very fundamental one!

Explaining the Different Types of Data

There are different types of data that your organisation can collect. These tend to be broken down for convenience into the following categories:

- Unstructured Data
- Semi-Structured Data
- Structured Data

Let's look at what each one of these data types entails.

What Is Unstructured Data?

Typically, when we talk about unstructured data, we're actually talking about data that is highly complex in structure and therefore difficult to analyse. (Strictly speaking, this type of data still has structure and rules that determine that structure, but the structure is more complex and more difficult to interrogate.) The easiest examples are video files (whether a movie in mp4 or mkv format, or a streaming video feed from a security camera), audio files like mp3s and image files like jpegs and gifs. Unstructured data also embraces the category of rich documents like PDFs, Microsoft Word documents, spreadsheet files, and so on. Digital medical scans and the like also fall into this category.

Usually, the file structure of one of these files also contains its own metadata. For example, an image file often records the date and time the image was created, a location stamp, and an identifier of the device which created the image (whether a digital camera or smartphone). Mp3 files contain metadata on the bitrate of the recording, whether it's in stereo and so on. The metadata is used by software to help "replay" or reconstruct the file into something you can watch or listen to.

The really interesting stuff about unstructured data — which has only just become a reality due to the exponential increase in computing power (and lower cost) — is the ability to run analytics over large sets of unstructured data.

Machines have "learned" (i.e. been programmed in complex ways which can then self-modify) to detect meaning (remember our discussion on "data vs information") from the 1's and 0's that make up unstructured data. Voice-to-text is an example of this and is available on any smartphone. Once upon a time, security cameras would feed images to a massive array of television screens, which human security guards would have to watch to detect if someone was trying to break in or climb a fence into an area they should not be. Today, artificial intelligence can monitor an almost unlimited number of live video feeds and detect among the 1's and 0's the kinds of signals and patterns that indicate that someone is in an area or trying to break into an area they should not be. The AI can then generate an alert, which might require one guard to be on standby, rather than the 20 or 30 who might have been needed before to monitor all those screens. (As I mentioned before, one result of this is that the other 29 security guards are freed up from a relatively boring job to go and do something more challenging and fulfilling, which probably also adds more value to the economy than sitting in a room watching CCTV live streams.)

So we now have tools, often powered by AI, that allow us to parse through this unstructured data. Those tools will become even more valuable in the future since the bulk of the data we generate falls into this unstructured category.

What Is Semi-Structured Data?

Semi-Structured data lies in the middle — there is some easily detectable structure in the data, but not all of it is simple rows and columns (which would make it structured data).

The most common types of semi-structured data that we typically encounter include emails, interface files, tweets, and other social media entries (which can be accessed as "objects" via an API), java-based objects like json, and XML.

A typical characteristic of semi-structured data is that you don't always know how many fields or attributes will be populated, but if you have the "schema" or mapping (this is our metadata), then you can know all of the possible fields that can exist.

Emails are a good example. There may or may not be a subject, there may be one or many recipients, there may be zero or several attachments, and the text of the email might be short or long.

Another example is a Twitter "tweet". If you capture it for analysis, it can have anything from 30-odd attributes up to a couple of hundred. There are some mandatory attributes, like the person tweeting, the content of the tweet, date and time, and so on. Then there are optional attributes — the person may have their location hidden, so this data won't be included; in other cases, you can get the location (latitude and longitude) from where the tweet originated. Missing or optional fields like this are hard for a traditional database to handle (because it's designed for structured data) but formats like json and XML can handle this easily.

Running analytics over this kind of data is a bit tricky because of these optional and variable attributes, but it can be done. For example, with enough tweets, you can run sentiment analysis on a certain subject, identifying how people feel (positive or negative) about a particular subject in different countries, regions, states, and cities.

The potential for these kinds of insights makes working with semi-structured data a very interesting field.

What Is Structured Data?

Structured data is the easiest type of data to work with. Structured data refers to data that is easily stored as rows and columns, perhaps in a database or even in a spreadsheet (while the spreadsheet file itself might be semi-structured data, the data in the rows and columns is structured data). Each row is an instance or record of a particular object or event, and each column or field contains information about that object or event. The same format or structure applies to every record, and every record has the same fields (although some might be empty).

It's that rigidly enforced structure that separates structured data from the other two types. You may have a database that stores names, phone numbers, dates of birth, and postcodes. Each column or field often has rules applied to help ensure that only the right kind of data can be entered. For example, you won't be able to store a name in a date-of-birth field because that field requires date information in a specific format.

Let's take an airline's reservation system as an example. In this case, the airline collects very specific data from its customers when they book a ticket. This data is captured in a database or RDBMS. You may have one table that stores the customer record itself, and this may link to other tables related to the flight that they've booked. All of this is easy to search through and that search will deliver a "report" of sorts, for example, verifying information about the customer when they arrive to check in. The airport can instantly see the customer's details and details about their flight reservation by calling up the data. And if they want to delve into the data on a wider level, such as finding out which destinations prove most popular, they can easily run a query on the collected data because it has this known and non-violable structure.

We can further divide structured data into two categories depending on what it's used for in the business:

- Operational Systems
- Analytical Systems

The airline system described above is a good example of an operational system. In this case, the data is used as part of a transaction like a purchase or a flight. A student record management system or a bank's credit card billing systems are also good examples. The data is created and used as part of a business process (remember the BIAT model) such as a purchase or a course enrolment.

Analytical systems are databases that are used to look at huge volumes of data over time, to try and understand (analyse) what is happening in the business, identify trends, and make decisions. Often this is done in a dedicated repository like a data warehouse. The data warehouse is a highly structured database that brings together data from many different systems so that it can be analysed in one place.

To illustrate the usefulness of data warehouses, let's look at one of the industries that were the first to use them: banks. Typically, a bank will have a system (an operational system) that manages applications for and transactions on credit cards. There will be a different operational system for mortgages, a different one for personal loans, and another one for margin lending. To get a view of a customer's total borrowings would mean running queries in all these systems and then adding up the results. Or... we could bring the data from all of these systems together in a way that makes it easy to run a single query over all the combined data. That's a data warehouse, an example of an analytical system.

Interestingly, unstructured and semi-structured data are now also being captured for analytics. This kind of data is not easily handled in data warehouses, as they were designed with structured data in mind. So, we have to use a different kind of storage, sometimes called a "data lake", to accommodate these kinds of data. There will be more to say about this later.

Taking Data from Operational to Analytics the Traditional Way

Let's drill into this concept of taking data from operational to analytics in more detail. The following is a model for the traditional data reporting and analytics, which we can refer to as the data warehousing model:

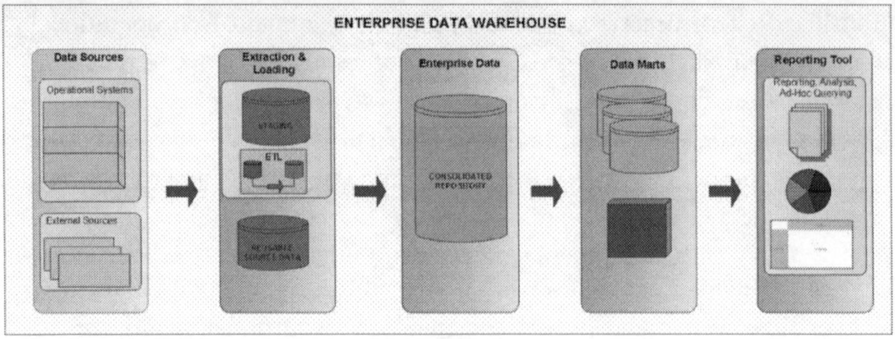

At the first step, our operational systems act as a source of data for our analytical ones. That leads us to the natural question of why we don't just run analysis on our operational systems. There are three key reasons:

1. Running sometimes complex queries on an operational system can create a very high workload on that system and slow down its primary function of processing transactions. (This has become less of a challenge now, as it's easier to scale modern apps than it was to scale mainframes in the past, but it's still an important consideration.)
2. The complex data models used in operational systems, especially transactional systems, don't lend themselves well to reporting and analytics. Therefore, we need to build a more user-friendly data model for analysing our data. That's where the analytical systems come in.
3. You often need to combine data from multiple systems to run your analysis. The goal with an analytics system is to put all of that data in one place so you're not constantly pulling from different systems to create reports.

However, there are also challenges with the model that I've shared above. This is a traditional data warehouse structure that ingests data from multiple systems and merges and transforms it into a single enterprise data model. In theory, this sounds like an excellent way to provide for analysis of all the data in the organisation. But in practice, this model requires an enormous amount of work to create. They're expensive to develop and maintain, and if you buy a "pre-packaged" analytical data model from a vendor, it may not answer the questions that are specific to your organisation.

Typically, this enterprise data system will come from an external vendor. I recall an example of a bank in Singapore that shows how costly this traditional model can be. It spent $2 million to purchase an analytical data model (in this case, it was IBM's Financial Services Logical Data Model or FSLDM). Once the model was in place, the bank spent an additional $10 million trying to code and transform its data from all the source systems (like credit card systems, mortgage systems, margin lending, retail banking, and so on) to fit the new model. Following this, of course this single enterprise data model (in the "consolidated repository" area of the diagram above) was so complex that it wasn't usable by the average person. So, they needed to invest even more money into creating simplified subsets of the model to make it usable by creating what we call "data marts" for each of its divisions. Sales, product performance, customer experience, and so on all need access to this data store in different ways.

This entire process took three years to complete. And once done, the bank had a data warehouse that it could use to answer questions from three years ago. It wasn't aligned with the questions the bank needed answers to today!

We can see that following the traditional model for moving from operational to analytics is costly and time-consuming. We can also see the limitations of the data warehousing concept with the example of the bank that I just shared. However, this doesn't mean that the entire concept of creating analytical systems doesn't work. Instead, we just need a better way of doing it.

Before we do that, however, I'd like to introduce one more key concept...

CHAPTER 7

Tech Debt and Data Governance Debt

Technical debt or "tech debt" accrues when organisations opt for a quick fix to an immediate IT problem. It solves the problem cheaply in the short term, but it comes at the cost of not fitting in properly with other systems or with the enterprise architecture and standards (because to do that would have cost more and taken longer). It usually also creates a need for ongoing manual intervention and maintenance, hence ongoing cost.

I remember working with a company where a classic example of tech debt was incurred. The renewals team wanted to roll out a "next best offer" system so that any customer who was nearing the end of their phone contract could be given an offer to re-sign automatically, with an appropriate contract and incentives depending on their spend, usage, overseas travel, and so on. They purchased a software product and got it into production so that emails with the right offers went out automatically. Because they felt they were losing revenue opportunities, they fast-tracked the project and ignored the potential to integrate the system with their customer call centre and their web development team. But their "omni-channel strategy" was that customers should receive the same offers whether they logged into the customer portal or website, by email, or if they called the customer service centre. Integrating this new system with the website and the call centre after it was already in production was many times more expensive than it would have been if the integration had been scoped as part of the original project — and took longer as well. By fast-tracking the first project, they accumulated expensive tech debt for later projects to fix.

Of course, they did gain an early benefit from rolling out the project faster. But the extra revenue from those renewals was nowhere near to making up for the cost of the later projects.

The more you save on quick fix after quick fix, the more you have to fix up and do properly further down the track. At some future time, those organisations will have to spend a lot more to catch up on the areas in which they took shortcuts — properly integrating with other systems, investing to properly automate manual processes, and so on. This is how organisations accumulate "tech debt". They save money in the short term on a technical solution but accumulate a much greater cost which has to be addressed in the future.

Data Governance Debt Is the New Tech Debt!

We see a similar thing happening with data governance. An organisation may opt to save money in the short term by skimping on data governance, since the value may not be immediately evident.

For example, the organisation may choose not to create a data dictionary. Or, they may not "get around to" updating the Business Glossary when business rules or the regulatory environment changes. This short-term saving leads to an accumulation of "data governance debt". Eventually, those issues will need to get fixed, whether because of legislative requirements or because the data becomes untrustworthy. And thus, the debt accumulated early on has to be repaid and costs much more to fix later than it would have cost to do properly in the first place. "A stitch in time saves nine" also applies to data governance!

A perfect example of Data Governance debt comes from a large telecommunications company that I worked with. When I started, they'd arrived at the conclusion that the data warehouse on which they spent millions of dollars every year was not able to be trusted as a source of truth. Analysts would run standard reports using the data warehouse but then also run extracts from other systems that they trusted more to cross-check the data warehouse reports. Obviously, this was hugely inefficient.

The specific problem here was that the documentation around the entities, attributes, and metrics in the data warehouse was never done properly. Even worse, the consultancy that set up the data warehouse now wanted a six-figure sum to come in and do proper documentation on the data warehouse they had set up some years before.

During a meeting with that consultancy, I asked the important question: "Why didn't you guys document all this properly in the first place when you built the thing?"

They told me that the CIO of the telco at the time needed to reduce the cost of the project. The original engagement included a full data dictionary, data lineage, and metadata management as part of the project. But the CIO de-scoped it to get the price down, so it was never done.

This decision incurred an enormous amount of "data governance debt" for this company. And what's worse, the decision also massively reduced the value of the data in the data warehouse, significantly hindered their ability to become data-driven, and reduced their ability to compete in the market.

Proper data governance ensures that this debt never grows in the first place. In other words, good data governance requires anticipation, foresight, and long-term cost-benefit analysis. The data infrastructure and foundation that you build today will be essential in the near and distant future.

SECTION 3

The How of Data Governance

CHAPTER 8

Organisational Structure (and the Rise of the Chief Data Officer)

In 2019, Big Data and business strategy consulting firm NewVantage published an interesting study. Their goal was to understand how well organisations actually execute on the vague concepts of being "data-driven", "analytics-enabled", and "AI-empowered". They didn't survey small or immature organisations as part of this study; they were looking at how the top players in the business world perform. These are data-mature firms that spend significant sums on their data processes and project a reputation of being at the forefront of data management and innovation. Examples of those surveyed include the likes of MasterCard, Bank of America, General Electric, and Johnson & Johnson, to name a few.

In total, NewVantage spoke to over 60 firms.

Most of those surveyed had a budget of over US$50 million per year to support their ambitions of being "data-driven" and were increasing this spend year-on-year by 37% — not a trivial investment and not a trivial growth curve. Furthermore, 88% considered their investment in analytics as being an "urgent" priority for their business, with 75% of the organisations anticipating disruption due to the data revolution that I've already talked about.

So, it sounds like the major players are all making big moves. Surely, this should mean that they're well-prepared in terms of managing and leveraging their data in the coming years.

Perhaps surprisingly, that isn't the case. Despite all of the good intentions and impressive budgets, we saw some familiar feedback when the organisations talked about their data readiness:

- 72% said that they had not been able to successfully create "a data culture".

- 69% stated that they could not describe themselves as a data-driven organisation.
- 53% believed they were not yet treating data as a valuable business asset.
- 52% admitted they were not "competing on analytics" (to use the term coined by thought leader Thomas H. Davenport over 20 years ago).

Considering that these organisations are massive, with huge budgets, and portray themselves as being at the forefront of innovation and data exploitation, it makes for a depressing reading — even more so as we're entering this new period of data-driven innovation and efficiency. It behoves us therefore to look closely at the reasons why these organisations still aren't being successful in their initiatives to become data-driven and to compete on analytics.

It comes down to one of the key ingredients for success in data governance: the culture in these businesses and how their organisational structures interact with this culture. We can see recognition of this in the survey, with 93% of those questioned pointing to "people and processes" as the major hindrance to becoming data-driven.

Interestingly, 40% of those surveyed pointed to a lack of organisational alignment as the key issue here. And yet, only 24% identified cultural resistance to the concept of enhanced data governance.

What does this tell us?

We're not seeing active resistance to data governance. Instead, we're seeing organisations struggle to get traction and adoption at the cultural level right from the very beginning, which means their data governance initiatives struggle to make an impact.

In this chapter, we'll examine the structure, roles, and skill sets needed for a successful data governance initiative. In particular, we're going to examine three key roles in your organisation that you need to fill, starting with the Chief Data Officer.

What Is a Chief Data Officer (CDO)?

I first came across the named role of Chief Data Officer in 2004. In Australia, Telstra was the first company to appoint someone to this role — quite an innovative move for its time. The trigger was the realisation that they could not transform or compete (with the newly minted "Aussat", later to become Optus) unless they learned how to treat and value data as an asset at the enterprise level. To do this, they needed someone at the C level, not part of the IT org chart, who could advocate across both the business and IT to fix broken data processes and uplift the data maturity of the organisation. Following an emerging trend in the United States, they appointed Australia's first Chief Data Officer.

Sadly, after a few years of real innovation and progress, a newly appointed CEO saw the term "Data" in the CDO title and reshuffled the org chart, pushing the CDO into the IT organisation reporting to his newly appointed CIO, back inside the IT org chart. Needless to say, it all went pear-shaped after that.

So, what is a Chief Data Officer meant to be?

One of the simplest and best definitions comes from Sunil Soares, author and founder of Information Asset, who defines the Chief Data Officer role as follows:

"The Chief Data Officer is a C-level role with overall accountability for data as an enterprise asset."

The emergence of this role is a reflection of how data has shifted from being a by-product of a business process to being central to the business process itself. Unfortunately, in many organisations, the culture that grew up with data as a by-product has not evolved to the understanding of data as an enterprise asset and a driver of competitive advantage and innovation — despite grand platitudes to the contrary. The proof is in the survey we cited earlier in this chapter.

The diagram below might help to illustrate what has shifted, and why we have seen the emergence of the Chief Data Officer. Recall the BIAT model we referred to earlier:

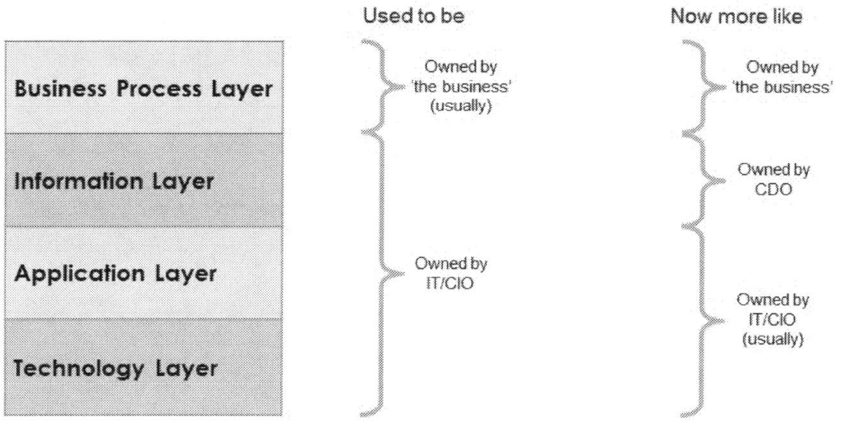

What's ironic is that the Chief Data Officer is actually responsible for the lifecycle of *information* (which you would think would belong to the Chief *Information* Officer). On the other hand, the Chief Information Officer is actually looking after the *data* (and the systems that support it) — which sounds more like it should be the remit of the Chief *Data* Officer!

This likely stems from how we used to call all data systems "Information Systems", even when they weren't technically handling information. In the past, we saw data as the by-product of doing business. But today, data is the lifeblood of an organisation, and the ability to use it well is a critical competitive differentiator.

The role of the CDO is actually a relatively new one and has emerged in response to the need to treat data as an asset at the enterprise level. Hence the need for a voice in the C-suite to advocate for efficiencies and best practices in a coordinated way across the full information lifecycle and across the whole of the organisation — not just IT.

A CDO works in a business-leading role — and carries a remit across the entire organisation — rather than being part of the IT organisation and accountable to that reporting line. For data and information to be treated as an enterprise asset, a CDO must take on a transformational role in which they foster cultural change across the whole organisation, to uplift data maturity across business processes, rather than permitting data to continue to be seen as the remit of IT.

To achieve this, they and their team need to advocate at all levels of the org chart for changes to ways of treating data that allow the full value of the data to be exploited. We will dive into how this plays out when we look at the Data Governance Framework.

The most important thing is that data governance falls under the remit of the CDO.

To Whom Should the CDO Report?

Next, we come to the question of who a CDO should report to. In the case of Telstra back in 2004, the CDO reported directly to the CEO, a move which was made to deliberately elevate the role of the CDO and distinguish it from traditional IT. When the new CEO moved the CDO role under his new CIO — making it a technical and IT role — the wheels fell off.

That's not to say that a CDO can never report to a CIO, it will depend on the nature of the organisation. The truth is that the CDO may report to any number of C-level executives, depending on their specific function in the organisation.

Let's explore some of these options and the pros and cons of each.

The CDO may report directly to the CEO. This approach elevates the CDO outside and above the IT function and provides the right level of sponsorship for true cultural change. This approach is recommended where cultural transformation initiatives around data have failed in the past. The danger here is that the CEO is the only level that issues facing the CDO can be escalated to, and a CEO probably doesn't want to be brought into turf wars between C-level executives over who owns which part of the organisational strategy and its execution. However, with visible sponsorship from and a direct reporting line to the CEO, this approach sets the CDO initiative up with the highest chance of successfully driving transformation and cultural change.

I've also seen the CDO reporting to the Chief Financial Officer (CFO). This is often the case where the finance function is one of the most powerful in the organisation and is the business owner of the most critical data assets (think banks, insurance, and other financial services). The danger of this approach is that functions outside of finance may not get the attention they need to help get their data house in order (marketing might be an example).

The CDO can report to the CIO; the strength of this approach is that it should foster a strong alignment between the technology strategy and execution and the data strategy and execution. The weakness or danger is that data will continue to be seen as a technology issue, not a business issue, and cultural change does not take root outside of IT. This structure is perhaps the most common due to the fact that most organisations still see data as sitting in the IT domain and, for exactly the same reason, this is the approach most likely to fail.

In other cases, the CDO may report to the Chief Marketing Officer (CMO). For example, a digital marketing outfit may find that this structure works best for them, as Marketing owns the most critical data assets.

And again, in some cases, the CDO may report to the Chief Risk Officer, where data risk is seen as one of the most critical elements of data governance. (This needs a pretty data-savvy CRO to work well, or one with enough clout to back the CDO in areas of data management outside of just data risk management).

Finally, the CDO may report to the Chief Operations Officer (COO). This provides a very strong model as the COO is often seen as more neutral and objective. Also, the COO role is about process optimisation and value optimisation, and the CDO goals align well with these values.

The decision comes down to figuring out how the CDO can best execute their goal of bringing about cultural change. Work out where that change needs to emanate from — be it finance, marketing, or operations — and determine your reporting structure from there.

Critical Success Factors for the CDO

There are some particular success factors for a CDO which I'd like to highlight. In other words, you can tell whether or not a CDO is likely to be successful and whether the organisation is truly embracing a culture of data governance by these indicators. These are the key questions to ask in this department:

- Is the CDO's extended team big enough to have a contact point in every business function?
- Are the boundaries between IT/CIO and the CDO clear and agreed? (Or will the CIO or rogue minions try to undermine CDO initiatives during a perceived "turf war"?)
- Can the CDO purchase and implement software to support its mandate (e.g. a Data Catalog, an MDM platform, Data Quality tools, etc.)?
- Will the CDO have representation in key strategic business processes such as the overall organisational strategy and planning, the budget cycle, the project management office (PMO), and other forums?
- Will the CDO be able to pause or red-light a project that has not addressed data governance issues?
    - For example, will the CDO have the ability to halt a project if data security or data privacy have not been properly addressed? This kind of authority would have saved Westpac from putting a defective application into production and allowing money laundering and child exploitation to take place. The cost of a delayed project pales into insignificance against the fines that were later levied for this breach!
- Will policies, standards, and processes established by the CDO have teeth?

If any of these factors are missing, it's worth revisiting whether the organisation is serious about managing data as an enterprise asset, reaping the benefits of a data-driven culture, and addressing the risks of not managing data properly.

Introducing the Data Steward

The Data Steward is a critical supporting role for the CDO and the Chief Data Office. Often, the data steward is a part-time role that is awarded in addition to someone's regular day job within a line of business. They will generally be a subject matter expert within their business unit who works with data on a day-to-day basis, such as a data analyst, business analyst, BI developer or data engineer, or indeed any other information worker or subject matter expert within a line of business. Often you can identify the ideal data steward as they are the "go-to" person for anyone with questions about data within that part of the organisation. In fact, they are usually acting as an unappointed data steward anyway, as they see the value in elevation data management processes and tend to be an evangelist for data best practices.

A Data Steward is part influencer and part knowledge curator. Their responsibilities extend to advocating for improved data quality and governance in all business activities and projects that impact their area of "data remit". For example, you may have a Data Steward in marketing who focuses on ensuring the marketing team receives high-quality data for their activities. A Data Steward also acts as a subject matter expert and point of contact for data-related queries and issues in their area of work and expertise. Furthermore, they help to define the roadmap for how the larger data strategy needs to support their function and area of remit.

Finally, a Data Steward identifies opportunities for data quality uplift in existing processes and calls out opportunities for new processes to uplift the quality, usability, and availability of the data in their area. To the extent that they are involved in new projects, they also identify these opportunities at the project definition phase and advocate for these improvements to be baked into the project.

But perhaps most importantly, your Data Stewards are advocates for change. They are the people who'll spearhead cultural transformation within their departments, thus helping to create a more data-driven organisation.

One final note: when appointing a data steward to represent the Chief Data Office and to drive the CDO's mandate into their part of the organisation, the appointment has to be carefully positioned and managed. The last thing a busy person needs is more responsibility piled onto their already full workload. We will look later in this chapter at the two parallel org structures into which a data steward belongs and reports: their "day job" org chart and the CDO org chart. It is critical to the success of this role that the data steward's "day job" manager understands and is supportive of the CDO's initiatives and of the role that their team member will play as a part-time data steward.

The Data Steward Role Summarised

The Data Steward is:

1. A Knowledge Management function: to accumulate, collate and disseminate knowledge of the datasets (and data-related systems and processes) under their remit (especially by using the Knowledge Management systems such as the data catalog or a Business Glossary and Dataset Dictionary), and

2. An advocacy function: to advocate for change (process improvements) as required to advance uplift in:
   - Data quality
   - Data usability (including metadata and data lineage)
   - Data availability (to the right people for the right uses — which includes privacy and ethical considerations)
   - And ultimately, the value that can be derived from trusted data.

How do they do this?

They do this in collaboration with business process owners and system owners:

1. As part of BAU (as they go about their 'day job').
2. In response to data quality or other data-related incidents which highlight the need for change to processes or other governance-related matters.
3. As stakeholders in projects that impact or make use of the data in their remit.

The Complex Environment of the Data Steward

The Data Steward will find themselves grappling with a number of realities (in addition to their "day job" as an analyst, etc.).

Data Knowledge Management

Knowledge Management, and knowing how to facilitate the capture and preservation of knowledge around the datasets within their remit, is a critical success factor for a Data Steward. This includes ensuring the accuracy and "up-to-dateness" of the Data Dictionary, Business Glossary and other Knowledge Management assets as well as acting as the SME in meetings relating to decisions on data usage. They will have an awareness of the lineage of the data they steward and understand what happens to the data on its journey to the point where their business unit uses it. They will have an awareness of the sensitivity of various datasets in their remit and therefore the security and access requirements that result from that. This includes knowledge of how the organisation has chosen to apply legal, contractual, and ethical principles to data availability. They will be aware of existing and potential data quality issues, and what that means for how the data can and should be used.

Data Issue Management

A "data issue" occurs when some impactful event happens that results from poor data or poor data practices. For the Data Steward, they need to be thinking about data issues in two ways:

Reactive — what are the processes by which we respond to a data issue, such as data quality or data security/access issue? How is an issue raised, managed, resolved, and then learned from?

Proactive — anticipating potential data quality and data governance issues before they arise in production or in the real world and taking steps to see that the issue or event does not occur, and how to minimise the impact of the issue if/when it does occur.

Data Culture

Part of the Data Steward's role is really a "soft power" role of diplomatic influence and winsome persuasion — strategically highlighting the value of good data governance and data management processes.

By virtue of their role, the Data Steward will often have a seat at the table in project planning meetings, status meetings, root cause analysis meetings for issues that have occurred, BAU-type ongoing forums, and even strategic planning or review meetings. These forums provide an opportunity to call out the positive business impact of good data governance, and to clarify how it actually makes everyone's life easier in the long run even if it means a little extra work in the short term.

Project Influence (Not Management)

The Data Steward is usually not a project manager, but they will often be given the privilege of attending project meetings and having an influence on projects that are planned or underway relating to the datasets within their remit. In this capacity, they have the opportunity to influence the project to "get it right the first time" from a data governance perspective. Specifically, opportunities to improve data quality, data management processes, data knowledge retention and updating, data availability (to the right people, hence also data security), data usability and ultimately, data value frequently found in the planning, scoping and design phases of a project that are very cheap to implement early on but hugely expensive to fix after go-live. This is where the Data Steward can play an incredibly valuable role.

Data Risk Management (Including Risk Identification)

Again, this is in both the contexts of BAU and project development. As part of BAU, the Data Steward interacts with people while working with or providing access to or information about the datasets within their remit. Risks or risky processes can be uncovered along the way. These can then be documented and appropriately escalated through the Business Data Champion (a role we will introduce below) and/or CDO, to be added to the appropriate workflow to be addressed in due course — depending, of course, on the severity of the risk if it does occur and the likelihood of the risk actually occurring.

Turf Wars and Politics

Unfortunately, this is a reality that a Data Steward may sometimes be caught up in. Large organisations, in particular, can host side-show battles between egoistic executives bent on furthering their careers or reputations regardless of whether or not their path to glory aligns with the overall strategic success of the organisation. At other times, there are just tactical decisions made by executives for budgetary reasons outside the data steward's control, which might lead to a missed opportunity to avoid the data governance debt mentioned earlier.

There sometimes isn't very much a humble Data Steward can do. By having a lot of conversations with a variety of people across the organisation, they can occasionally get an idea of what's going on in the background — and knowledge is power. They just need to pick the battles that are worth fighting and where there is a chance of winning. In other cases, it's necessary to emotionally detach from the issue and just know that they have done what was within their remit so that they can "sleep the sleep of the righteous" while the political players undermine the value across the organisation!

The Qualities of a Good Data Steward

It's likely that you will promote from within when assigning the Data Steward role. Therefore it's important to know what to look for when searching for somebody who fits. Remember that this is a business-facing role rather than an IT-facing one. This means that technical skills are not the highest on the list of priorities. Instead, look for the following qualities in a potential Data Steward:

- The ability to listen well and understand the organisation's overall goals in relation to its data governance policy. A Data Steward must be willing to accept feedback at all levels of the organisation to achieve the best possible result.
- A collaborative nature is a key quality for a Data Steward to be successful. They must seek to contribute value to the team as they find opportunity.
- It helps to have a positive outlook on their work, which aids them in finding opportunities. This doesn't mean that they're incapable of pointing out problems. It simply means that they actively look for solutions upon identifying any problems.
- They're strong communicators who have the ability to promote the importance of good data governance throughout the organisation. They can articulate the value of data and why it's an asset to the area in which they work. Most importantly, they're capable of using these communication skills to obtain buy-in for data governance at all levels of the organisation.

- Data Stewards occasionally need to step up into a leadership role when nobody else seems capable of making a decision. They must do this with confidence, but not arrogance.
- Finally, Data Steward holds themselves accountable for the work that they do. They take responsibility for finding new ways to gain value from the datasets that fall into their remit.

If the CDO is the facilitator of change, the Data Stewards are the instruments that create that change. The qualities noted above should set a Data Steward up for success in helping implement the changes that are needed on the path to better data governance.

The Business Data Champion

The most successful organisations I have worked with, especially larger companies, find the need to appoint a layer of responsibility between the CDO and the Data Stewards. One title for this role (which I have borrowed from an Australian bank) is the Business Data Champion.

The role of a Data Steward can be a lonely one — they can easily feel like they're the only person who's actively advocating for data quality, data security, knowledge management, and all the other areas we've identified as part of data governance. It can feel like you're always wearing the "black hat" (referring to Edward de Bono's *Six Thinking Hats*) — finding the risks, highlighting extra work for everyone to ensure that the data is treated as a business asset, and that potential future data issues and risks are nipped in the bud. These challenges mean that the Data Steward can benefit significantly from having a coach and a mentor, helping them to operate effectively and sustainably over time. That's where the Business Data Champion comes in.

The Business Data Champion can be a full-time or part-time role depending on the number of Data Stewards they are supporting (between 3 and 8 seems optimal for a full-time Data Champion), the number and complexity of the data subject areas within their remit (which corresponds to the remits of their Data Stewards) and how far from ideal the current situation and processes are.

The Business Data Champion Role

The Business Data Champion:

- Acts as the escalation point for Data Stewards where conflicts or issues cannot be addressed or resolved at the Data Steward's level.
- Acts as a mentor and coach to multiple Data Stewards.
- Represents their subject areas and data repositories (per their Data Stewards) on project steering committees, data governance forums, and similar.
- Reports to the Data Governance Executive(s) who is/are responsible for the corresponding datasets and data repositories (could be multiple executives).

A high performing and passionate Data Steward is the perfect candidate for a Business Data Champion role if they have had time to grow into the next level. If you're just getting started with your data governance initiative, then the Business Data Champion themselves may need some training. If the knowledge and skill sets to train these Data Champions is not available within the organisation, external training is the fastest way to get them up-to-speed. Some options are laid out at the end of this book.

The Executive Data Owner

The final piece of the puzzle is an executive-level owner for each of the organisation's critical data elements (CDEs) or most critical data subject areas. In most organisations, this is a GM/VP level or EGM/SVP level leader. In a smaller or less complex organisation, the CDO may act as the Executive Data Owner for all key data assets.

The Executive Data Owner is a role that is expected in organisations that are supervised by APRA (per CPG 235) and the role comes with some specific legal and regulatory expectations. Even for organisations that are not APRA registered, it should be regarded as best practice to have this executive ownership structure for the organisation's key data assets.

The Executive Data Owner:

- is the executive business owner of the dataset(s) curated by the Data Steward(s) in their team.
- makes the final call on agreed definitions for key terms within their domain such as "customer", "revenue", and so on.
- reviews and approves key reference data and hierarchies (MDM) related to their data domain (or delegates as appropriate).
- makes or approves decisions around data privacy, data usage and access, data sharing, data commercialisation and data desensitisation, and data retention and disposal.
- consults and collaborates with peer Executive Data Owners across domains to resolve cross-domain data issues or questions.
- reviews the level of data quality that is required for the data to be fit for any particular purpose and sponsors any data quality uplift initiatives that might be needed from time to time (typically, they would review and approve/query/refine the recommendations from their Data Stewards).
- acts as escalation and resolution point for data-related issues that cannot be resolved by the Data Stewards or Business Data Champions in the relevant forums and committees.
- provides input to the Data Governance Steering Committee on software solutions, policies, or Regulatory Requirements that impact their data domain.

What It Looks like

The kind of organisational structure I'm talking about here is illustrated in the example below.

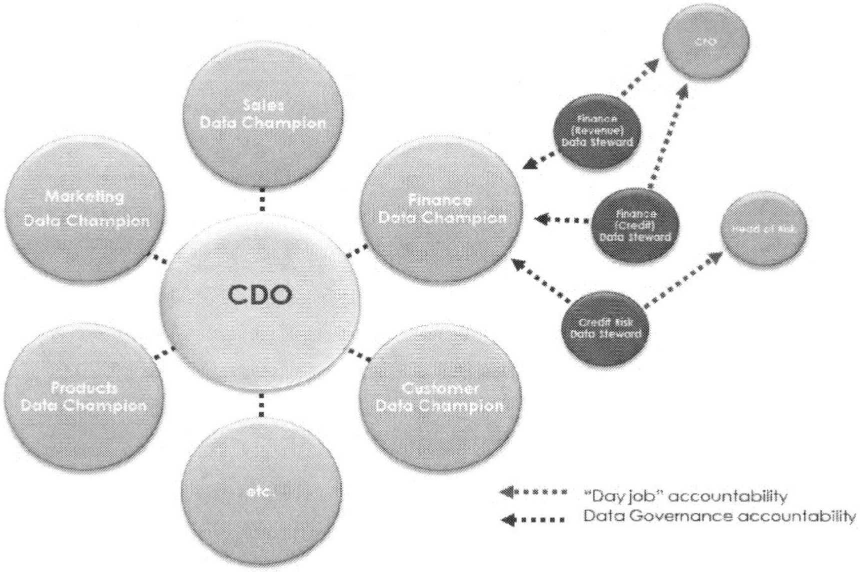

Other Models

Centralised Model

In a centralised model, all data governance functions and responsibilities sit within a centralised team who provides consulting and guidance across the organisation both proactively and in response to requests. This approach can work well in smaller, less complex organisations but doesn't scale well.

Decentralised Model

This is where pockets or silos of data governance practice occur throughout the organisation. This can work well when the organisation is large enough that appropriate autonomy of the data governance function at some level of distribution provides greater efficiency and effectiveness (especially when geographically distributed across different legal jurisdictions). This does not negate the need for a group level CDO and centre of excellence to provide overarching guidelines for the Enterprise, with the flexibility of interpretation and implementation at the local level.

Governing Bodies

The CDO may chair one or more councils or committees that he/she deems necessary to bring various data stakeholders together from across the organisation to make corporate decisions and the executive level — such as a Data Governance Steering Committee. The titles and agendas of these governing bodies will vary from organisation to organisation or may be deemed not to be required at all if issues can be resolved by taking a tactical or "swarm" approach to assemble the right stakeholders to deal with an issue and then disband the temporary working committee. In larger or more risk-exposed or regulated organisations, there may be a separate structure for risk management that parallels the structure for data governance; in other cases, data risk remains under the data governance umbrella and the risk org structure may be the same people as the data governance org structure. Efficiency and effectiveness are the guidelines that will steer towards the best approach.

What Next?

Getting the organisational structure, the lines of reporting and accountability, the role descriptions and demarcations of responsibility right is a crucial step in successfully rolling out a data governance program, or indeed improving an existing one. These role descriptions need to be informed by the existing org chart, the Data Governance Framework to be implemented and, to some extent, the technologies that are involved.

The rise of the CDO in the 21st century hasn't occurred by chance. It's an evolution in organisational structure brought on by the change in how we view and use data. Developing this role in your organisation, as well as furnishing the CDO with Business Data Champions and Data Stewards, is a key part of creating a data-driven organisation.

With them in place, you're in a position to implement your Data Governance Framework.

CHAPTER 9

Creating Your Methodology and Processes: The Data Governance Framework

In Chapter Six, we discussed the concept of "tech debt". It's an increasingly important factor when making technology decisions in any IT department: we need to be very careful when considering low-cost quick fixes over more costly but strategic solutions that are more sustainable. This is because the cost to fix a "quick fix" is usually more than the cost of doing it properly in the first place. To avoid tech debt, one must consider how any new form of tech will fit into an organisation's long-term roadmap.

In a similar way, organisations often confront opportunities to incur or avoid "data governance debt". Like the tech debt, the data governance debt causes long-term problems that are more expensive to resolve further down the line. The Data Governance Framework provides a set of principles, guidelines, and processes to minimise data governance debt by getting data governance right at the earliest opportunities.

What Is the Data Governance Framework?

Think of the Data Governance Framework as the methodology and processes you put in place to manage your data. The Data Governance Framework breaks down the component dimensions of what needs to be taken into account when implementing a Data Governance initiative.

The Data Governance Framework has eight components:

1. Metadata and Metadata management
2. Data Lineage
3. Data Usability and Critical Assets to Support Usability
4. Data Availability and Access
5. Data Security
6. Data Risk

7. Master Data Management
8. Data Quality and Controls

In this chapter, I will run through each of these components in detail and unpack how to leverage these concepts as part of your approach to data governance.

Component #1 – Metadata and Metadata Management

The best way to define metadata is that it's the data that you hold about your data. Or more correctly, it's the information you have about your data assets and datasets, that provides the context that allows you to turn data into information.

Metadata may include any of the following:

- Descriptions relating to what the data is and what it means (which, of course, includes information that tells you how it can be, should be, or shouldn't be used);
- Information about the data format and its origins (where it came from and how it was captured);
- Any logic, filters, transformations, or business rules that have been applied to the dataset (strictly speaking, this is data lineage; however, the lineage of the data is also part of its metadata);
- Any restrictions or caveats in place that curtail how the data can be used (for example, some data may be classed as personally identifiable information or PII, and there are legal constraints on how this can be shared and used);
- Attributes related to the quality and usefulness of the data. For example, your metadata may help you see that a dataset is only relevant for a short period of time, such as location data of a person, which is out of date very quickly and hence no longer useful for immediate action (although the trends over time of the location and movement of the person might yield other patterns and insights).

As a form of data itself, metadata also has attributes that can be tracked. These include (for example) how complete and up-to-date the metadata is. Metadata is often retained, updated, and shared via a data catalog platform.

The data steward for each dataset is responsible to update and enrich the metadata for the datasets within their remit.

Metadata can be human-facing — readable in English, or machine-facing — readable by a process, API or application that can read the metadata in order to correctly use or process the data — e.g. in an XML schema (which was discussed in Chapter 6), the metadata is embedded along with the data, and the application that reads the XML uses the metadata to know how to process the data itself.

It is helpful to divide metadata into three categories, each of which differs in source/origin and usage:

Type #1 – Technical Metadata

Sometimes referred to as structural metadata, this is information about the characteristics of the data itself. For example, this type of metadata defines if a field contains text (and how long the text can be), a date or date-time stamp, or a number (and how many decimal places it can store). An example from a relational database appears below.

Results Explain **Describe** Saved SQL History

Object Type **TABLE** Object **BOOKS**

Table	Column	Data Type	Length	Precision	Scale	Primary Key	Nullable	Default	Comment
BOOKS	BOOK_ID	Varchar2	4	-	-	1	-	-	-
	ISBN_10	Varchar2	10	-	-	-	✓	-	-
	ISBN_13	Varchar2	13	-	-	-	✓	-	-
	TITLE	Varchar2	50	-	-	-	✓	-	-
	CATEGORY	Varchar2	15	-	-	-	✓	-	-
	PRICE	Number	-	6	2	-	✓	-	-
	NDING	Varchar2	1	-	-	-	✓	-	-
	PUB_DATE	Varchar2	4	-	-	-	✓	-	-
	AUTHOR_ID	Number	-	-	0	-	✓	-	-
	PUBLISHER_ID	Number	-	-	0	-	✓	-	-

1 - 10

Or in XML, the technical metadata defines what the data tells us, such as in this example:

```xml
<?xml version="1.0"?>
<quiz>
 <qanda seq="1">
  <question>
   Who was the forty-second
   president of the U.S.A.?
  </question>
  <answer>
   William Jefferson Clinton
  </answer>
 </qanda>
 <!-- Note: We need to add
  more questions later.-->
</quiz>
```

Here, the <question> and <answer> mark-ups are technical metadata that offers context for the records in the dataset.

Typically, this type of metadata falls under the IT department's remit. In fact, technical metadata is of no real interest to your end-user. A salesperson doesn't need to know if a field is CHAR or VARCHAR, for example. However, you need to have it in place to define the constraints that apply to your dataset as well as ensure its structural integrity. Thankfully, it's fairly easy to reverse-engineer technical metadata, as it's usually defined in some form in the system that holds the data or in the data structure itself. In a relational database, the "DESCRIBE" command or its equivalent can be used to get a dump of a table's technical metadata.

Data Catalog software typically scans all the technical metadata of every data source it can find to "catalog" all the data assets of the organisation as a first step towards enriching it with helpful and informative business metadata.

Type #2 – Business Metadata

This is information about what the data means in a business context. In other words, this metadata tells you how to use, understand, and analyse the data. Business metadata provides the context for the data that allows us to use it as information in a business process, to support decision-making — whether manual (a manager interprets the data and makes a decision) or automatic (an application or AI model reads the data and makes a decision). The importance of good quality business metadata will only increase as more and more routine business decisions are automated through process automation (which is a key aspect of the Fourth Industrial Revolution).

Typically, you'll capture this metadata in a data dictionary, which may be part of a data catalog platform. You may remember from the example I shared at the beginning of the chapter that the CIO of the telecommunications company decided not to pay for this metadata collation. This resulted in the data stored in the company's data warehouse offering less and less value over time, as the "tribal" knowledge or tacit knowledge (knowledge in people's heads) about the data gradually degraded.

Without access to business metadata, you're left in a position where you have to make assumptions about the meaning of the data. You may be able to discern some of its meaning from other information available, or you may be stuck with looking for the subject matter expert who knows how to correctly use the data. Otherwise, you can't safely and reliably use the data to support a decision.

Type #3 – Administrative Metadata

Administrative Metadata refers to information about how the data can and can't be used from a legal, ethical, or contractual perspective. Note that this is not the same as business metadata, which tells you what the data means and how you can derive value from it. Administrative metadata tells you what's possible with the data based on the constraints that exist, originating outside of the data itself.

Administrative metadata also includes data on who *has already accessed* a given dataset. It can be important from a legal perspective to retain records of who in your organisation has accessed or queried certain data. This information is usually stored in application or database logs and can be easily captured and retained with the right technology in place. We'll look at some interesting examples of this in law enforcement when we talk about data security and access.

One final note: some people refer to this kind of metadata as "operational metadata". This can be confusing as Operational Metadata is also used to refer to the monitoring of data transformation jobs (such as ETL jobs in a data warehouse) and which operations were successful and unsuccessful. Hence, the term Administrative Metadata is preferred.

Why Metadata Is so Important

If it isn't already clear from what I've just explained, metadata is the key to understanding your data. Some metadata provides information that's useful from a technical perspective. However, it's the business and administrative metadata that help you discern meaning from your data and, ultimately, inform you on how to use it in a business environment.

This is why it's so important to keep these kinds of metadata current across all stages of the data lifecycle. Any projects that generate new data assets or modify existing assets (whether modifying them in place or taking a copy and pushing the modified results to a new data repository) have a responsibility to record and update the business and administrative metadata related to these datasets — otherwise, it rapidly becomes "stale" and hence, practically unusable.

One final note to be aware of: depending on the context, what one organisation considers metadata may, for another organisation, actually be data and vice versa. You may remember the outcry that resulted from the revelation that America's National Security Agency was collecting data about the phone calls that Americans were making even when they were not under surveillance. The NSA dodged the illegality by saying that they were only collecting "metadata". In this case, the data itself was the voice content of the calls (which they were not collecting), while the metadata was the phone numbers, times, places, and duration of the calls. For a telecommunications company though, the phone numbers, time and duration of the call is the data, as it is the raw material for analytics such as pricing, network usage, trend analysis, next best offer, etc. I guess you could say, "One person's data is another person's metadata!"

Component #2 – Data Lineage

Data Lineage simply means where the data came from originally and anything that may have transformed or altered it on the way to where it is now.

As an example, the diagram below illustrates the concept of data lineage in the context of a data warehouse and how that affects the reports that are produced from it.

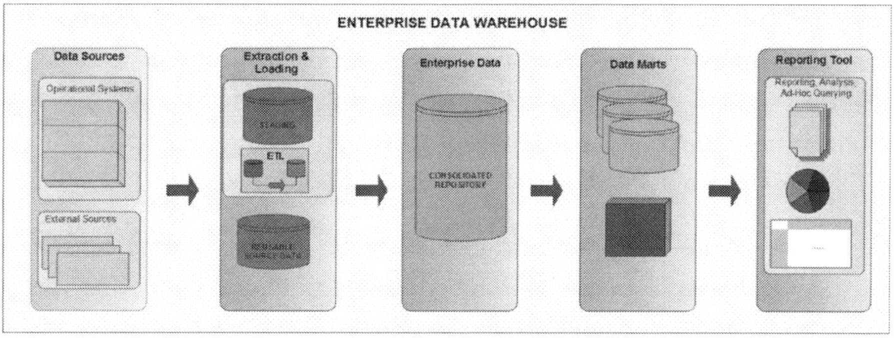

Each of the dark arrows is a point at which the data is transformed in some way. In the data warehouse, these transformations are "automatic" in the sense that the same transformation logic is run against every record in every dataset. By distinction, a manual transformation means someone has changed the data or applied some processing to it in a way that's not repeatable unless someone does the same thing again. Manual data transformations are far more "dangerous" in the sense that you can't easily trace back to determine what was done. Typically, this might be done in a spreadsheet as a final step before presenting a report to the management or an executive audience. This seems to occur frequently in most organisations (far too frequently, I would argue), but as soon as this is done, traceability of the data is broken — you can't prove where the final figures came from (unless all the formulas are still in the spreadsheet, and we can reverse-engineer the logic from them). These manual data transformations break the data lineage and often lead to mistrust in the reports that are provided. More importantly, if a regulatory authority such as ASIC or APRA requires an explanation as to how a certain figure was arrived at, you're on dangerous ground if the manual transformations that were applied can't be reconstructed and proven. Anyone can do anything in a spreadsheet, including making innocent mistakes, and there's no way to figure out what was done (unless as noted above, all the formulas are still in place and can be reverse-engineered).

Well-managed information on data lineage means you can prove where the numbers in any report come from.

We also need to keep in mind that we can think about data lineage in two ways: the technical data lineage and the business data lineage.

Technical data lineage refers to the code and the transformations that are applied at every stage along the data's journey. You can actually reverse-engineer the logic from the code if you have to, but this is an onerous job — it's much better if the code developer documents it all properly in the first place — and, critically, that anyone else who works on the code updates this documentation, as necessary.

However, a business user or an executive is not interested in the logic that's applied in the code. They need to understand the logic in plain English. We'll look at some good examples of this when we talk about Data Usability and the data dictionary.

Now that we've covered the importance of data lineage, let's look at how it works in a practical situation. Most reports are still sourced from a data warehouse of one kind or another, so we'll use the diagram above to unpack how data lineage works at each stage of the journey through those transformations.

The Data Sources

The source or origin of the data is a critical piece of the data lineage. It's where and how the organisation acquired the data in the first place. This could be through data entry into a system like Salesforce, SAP, or Oracle Financials; it could be from an API such as Facebook's data on who clicked on your ad; or it could be from an external source such as weather data from the Bureau of Meteorology, or demographic data from the Australian Bureau of Statistics.

How the data is captured is also important. For example, a call centre worker may have recorded data about a customer's income from the customer themselves. That's different from sourcing it from a copy of their tax return produced by the Tax Office. This means it's self-provided and not validated through an independent source, which is a constraint that may affect how the data can be used. These details need to be captured as metadata and form part of the data lineage story.

Finally, anything significant related to ethical and legal constraints on how the data can be used also needs to be captured at this stage. For example, we may have captured a customer's phone number, but they have opted to only receive marketing information via email. If this information is not recorded against the phone number, there are severe penalties under Australia's "Do Not Call" legislation for phoning them against their request. If the data is purchased from a data provider, there may be strict legal constraints on how it can be used, such as that the data cannot be shared outside the organisation that purchased it. Obviously, this is critical information that has to "follow" the data along its lineage journey — that is, people using the data in some "downstream" system need to be aware of these constraints to avoid using the data inappropriately.

The Landing/Staging or Data Lake Layer

Often, the data in this landing or staging area is untransformed from the source system, but it may retain more historical data than the source system itself, and hence can become a source of truth in its own right. Sometimes, this layer is called the source reference data or "SRD" layer. Similarly, this is often the approach taken for landing data in a Data Lake — it becomes a system of record for the source data which may not thereafter be retained anywhere else.

In the Data Vault methodology, this is referred to as the staging area and provides a persisted copy of the untransformed source data from which you then develop your data vault data models.

In a "big data" or IoT environment, you might already have aggregation, summarisation or filtering applied before keeping a copy of the data, due to the sheer volume that would otherwise need to be kept. This filtering might be applied using data streaming logic in a technology like Kafka, Nifi, StreamSets, Spark Streaming, and so on — or the data may be processed on the "edge" within or next to the device itself before summarised data is sent to the "centre" for historical retention and time series analysis.

People accessing the data at this stage need to know where it came from (the source), and then the logic, transformation, enrichment, filtering, or aggregation of the data that has been applied, if any.

It will also be useful to know how frequently the data gets loaded or refreshed; whether any data quality logic or error trapping has been applied, and what that means in terms of how the data can be used. This is particularly important when working with streaming data, as you usually can't go back and get another copy of the data from the source.

Going one step further, it is increasingly the case that organisations are not only acquiring streaming data into a data lake but also applying real-time processing to the streaming data based on machine learning and AI and triggering actions in real time (such as real-time alerts when a certain event occurs, or a certain threshold is exceeded, or an anomaly is detected). Someone needs to have the algorithms documented so that there is traceability around what is triggering the alert or the automated decision on the data as it streams in.

The Consolidated Data Repository

In a traditional data warehouse, this is the stage where data from various sources is consolidated into a single Enterprise Data Model. Tracking data lineage here typically revolves around recording the logic used to complete that consolidation.

For example, in a bank, there might be one system that manages information relating to credit cards, a different system that processes mortgages, another system for margin lending, another one for personal loans, and various other systems which might relate to credit facilities for businesses and so on.

The only way to understand all the products that a customer has with the bank — and specifically to calculate their total loan portfolio, their risk profile, or even just a "single view of the customer" for better marketing such as next best offer — is to bring the data from all of those systems together into a single data model. That's the CDR.

Now, often the complexity of doing this means that the bank will use a consultancy with experience in this area to build it for them. The consultancy may have done a similar thing previously for another bank, so they bring the same data model with them and tweak it for their new customer. (However, they usually price it at well above the "cost of tweaking".)

That's how the bank we referred to in Chapter 6 paid a few million dollars for a data model, many more millions to transform their data into it, and ended up with a data model that answered all the questions that a different bank was asking 3 years ago! (This is the worst-case scenario, not all data warehouse projects are that bad, but it's worth heeding the risk and adopting a business-driven approach to the requirements rather than an off-the-shelf approach.)

It's worth mentioning that this traditional approach to data warehousing is trending out of fashion, as it's highly complex (and therefore expensive) to implement and becomes inflexible as the business environment rapidly changes. The idea of the Enterprise Data Warehouse grew out of Bill Inmon's Corporate Information Factory — in fact, Bill Inmon is regarded as one of the fathers of the data warehouse (along with Ralph Kimball — they each took different approaches to the concept resulting in quite different disciplines which almost took on a kind of religious warfare between their adherents, but that's a conversation for another time). The idea was that if you brought all the information assets of the organisation together into a single data model, you could ask any question of that data model and get the answer. Sounds great in theory, but the reality is that it takes many months — often years — to acquire and transform all the data into one of these models, by which time the questions have changed!

As I've mentioned, a lot of money has been made by vendors and consultancies selling data models and then services to populate the data models. Usually, the promised data heaven is never arrived at.

The Data Vault approach addresses these limitations to some extent with a highly extrapolated data model that can cater better for changing business requirements and environments (but this is not something we can deep dive into in this book — there are plenty of great resources available now on this approach).

A couple of other approaches have emerged to address the limitations of the traditional data warehouse: the data lake and data virtualisation — both very different approaches to solving this problem (but again, not something we can dive into in this book).

The data lineage and metadata for this stage may be provided by the vendor (if the data model was purchased) — but even then, any customisations to the data model must be updated in the documentation.

So the CDR is meant to be the all-encompassing data model and single source of truth across any subject area in the organisation. However, it's usually so complex that you don't expose it to the end-user. Instead, out of the single source of truth, you develop subject-specific data marts.

Data Marts and Cubes

This is the layer that the majority of people will access to answer their business questions either through a dashboard, Business Intelligence (BI) tool, or for advanced users by writing queries directly against the database. A Data Mart usually consists of a set of tables that incorporate facts and dimensions for each business subject area (such as sales, finance, customer, and so on). A "cube" (also called OLAP or an OLAP cube) is a pre-built data mart with many calculations and date functions already compiled so that exploring the data becomes much faster — it's more of an interactive experience, whereas waiting for queries against a data mart to complete can interrupt the flow of thought when exploring the data in novel ways.

This is the layer where, having combined all the data together in the CDR, we then bring it out into subject-specific collections that are easier to navigate and query. (The CDR data model is usually so complex that most people are not able to use it practically.)

To use our example from earlier, this is where we would have a dimension like "customer" and a metric like "total loan value" — where the figures from the credit card, mortgage, margin lending, and personal loan systems are all brought together so we can see the bank's total exposure and risk with particular customers. This becomes more useful when you can analyse the data by city and state, by level of income, by profession, and so on. The idea is to be able to spot trends and anomalies and use this information to predict the market, reduce risk, bring a better product mix to certain types of customers, and so on.

It's in this layer and the next one (below) that business data lineage and the resulting business metadata become quite important. Regardless of what technical transformations and logic were applied as the data made its way to this stage, the business user simply needs to know what the data means and how to safely, legally, ethically, and accurately use it to make better decisions and do their job more efficiently and effectively.

This means that the most critical asset here is the data dictionary per data mart or cube. For example, in a "sales history" data mart, you need the appropriate definition of revenue, of gross profit, of margin, as well as the stores and store hierarchy (e.g. by region) and, perhaps, the salesperson involved in each sale, and their reporting lines (a sales rep for computers in a certain store may report to a product manager for the IT business rather than to the store manager). The data dictionary needs to explain the application of relevant business rules for how the data should be used; as an example, margin by salesperson might be defined differently from margin by store (e.g. one calculation might take rent into the cost base while the other calculation does not). If the user doesn't have this information, reporting may be less accurate or misleading; you could even be paying the sales reps the wrong commission, and then you're in breach of contract as an employer.

Reports and Dashboards

Sometimes there are calculations built into the reports or dashboards that people work with. The definitions of these calculations need to be available to those who are consuming the reports to ensure no one comes to an incorrect conclusion — and bad business decision — for not understanding the data they're looking at in a report.

Later on, we'll also look at the "Report Inventory" as an important component of knowledge management around this layer of data assets. In fact, it's generally considered best practice to have a formal process whereby report developers and power users can create reports for their own audiences, then if the report or dashboard is of ongoing or wider value, the report can be peer-reviewed, certified, and then published as an enterprise report (whereupon it gets an entry and description in the report inventory).

Some BI tools allow the creation of a mapping or "semantic" layer over a data mart, which makes these calculations reusable across multiple reports and dashboards. Again, the critical element is to ensure those who are using the tools or interpreting the reports understand exactly what they're working with, and thus are protected from misunderstandings, legal violations, and everything in between.

Component #3 – Data Usability and Critical Assets to Support Usability

While metadata and data lineage are important, their primary usefulness is to support data usability.

Think of this component as the safety rails that you put in place to ensure data gets used correctly. This is also where the concept of knowledge management becomes more important.

This component of the Data Governance Framework involves setting up the knowledge management standards, processes, and supporting technologies that enable this critical information to be captured, curated, and then provided to information workers. Some see this as a tedious exercise. But for me, it's one of the most exciting because it's here where you can make it easier for people to leverage data and do their jobs.

This is where you make huge strides in data efficiency, which pays off later on down the line.

Three key elements typically make up this component:

1. Business Glossary
2. Data Dictionary
3. Report Inventory

Each of these can be hosted in a standalone knowledge repository, but we're increasingly seeing these three artefacts come together into a "Data Catalog" that is a software product designed to host and publish all of these assets, plus information on technical, business, and administrative metadata and data lineage.

Regardless of the technology used, you've got to capture and manage this knowledge somehow!

The Business Glossary is a searchable reference that defines all of the key terms that are used across your organisation. It works in much the same way as the glossary that you might find in a book. The goal of the business glossary is to remove any potential for confusion about what a term means. That means the more thorough the entry, the more useful it is. Ideally, the Business Glossary should be hosted on a wiki-like platform, where more than one person can contribute and suggest improvements, which can then be approved by the relevant subject matter expert.

So, what does a good Business Glossary look like? Here are two examples of potential entries into a business glossary:

Consider "Gross Revenue is the amount of sales plus earnings from investments"

vs

"Gross Revenue is defined as the total volume (in dollars) of sales through all channels, exclusive of GST and any transaction charges collected by third parties (e.g. credit card fees), plus earnings (net of fees) from all investments. It is reported in AUD, with exchange rates for foreign earnings converted at the daily rate in the data warehouse (source: Bloomberg). See also EBITA, EBIT."

Which of these descriptions do you think is most useful from a data perspective?

Of course, it's the second one as this description of the term provides a lot of detail regarding the logic and rules that are involved. Aim to make every definition in your Business Glossary as detailed as possible, like the second example, to minimise ambiguity and ensure there's no room for confusion with the terms that are important to your organisation.

Here's another example from the Telco world:

"ARPU stands for Average Revenue Per User"

or

"ARPU stands for Average Revenue Per User per month and is calculated in any given context (e.g. total customer base, enterprise, SMB, consumer, or by-product or service) as the total gross revenue (see Gross Revenue) earned in that context (from that customer base or the component charged to a customer paying for a given product or service as part of their overall bill) divided by the number of customers/subscribers. Where more than one month of revenue is in view, divide by the number of months to arrive at the ARPU."

The Business Glossary needs to be thought of as a living document that will evolve over time as people detect ambiguities that need to be ironed out or nuances that need to be clarified. However, there should be a final authority for each entry, typically the Executive Data Owner, working with the Data Steward or Data Champion who is a subject matter expert in the area and who grasps the various enterprise-wide usages of the term.

The Dataset Dictionary

A Dataset Dictionary is much more than a list of tables. Rather, it's a dictionary of the key information that makes the datasets used in your organisation usable. Whereas the Business Glossary defines terms that are used throughout the organisation — regardless of where they might be found in various datasets — the Dataset Dictionary is a list of all the key data assets along with information on how to interpret and use each dataset. (I'm using the term Dataset Dictionary rather than Data Dictionary, as the latter term can be used to refer to a business glossary, depending on who's doing the talking.)

A good dataset dictionary will refer to the business glossary when an element in the dataset is an example of the definition in the business glossary. So, using our example above from the Telco, there may be datasets containing ARPU in many repositories: some for consumer data with consumer ARPU, some for the SMB segment and hence SMB ARPU, and so on. There might be a data mart for mobile device usage only and ARPU for that product. In each case, the people populating the dataset (the ETL developers or data engineers) need to ensure that their calculations reflect the Business Glossary definition of ARPU and then publish the definition in the dataset dictionary for the dataset they have created, aligned with the enterprise definition.

Note that a dataset could be a flat-file table in a data lake or could be a star schema that includes several well-structured tables built around an entity of interest. However, a dataset could also be a group of unstructured files, such as audio recordings.

Each dataset needs a Dataset Dictionary entry to help someone who wants to use the data to understand how to use that dataset appropriately.

The data dictionary will typically contain the following information about the dataset:

Notes on Origin or Source

This tells you where the dataset comes from and how often it is updated or refreshed.

Example:

"This dataset contains temperature and rainfall data for 15 collection points across the Hunter Valley wine-growing region. The collection points are expressed as longitude and latitude in degrees to 4 decimal places (not minutes and seconds). The data was sourced from the Bureau of Meteorology (BOM) records from January 2015 to December 2016. The dataset is not updated, and access to more recent data will need a new request to BOM."

Notes on Data Granularity

Granularity refers to the level of detail captured in the data in the dataset and how this affects how you might summarise or run calculations on the data.

Example:

"The dataset contains multiple readings per day for temperature — usually one reading per hour, but this is not consistent over the dataset for every

site. Rainfall readings are daily and are taken at 9 am and refer to total precipitation for the prior 24 hours, as recorded at that weather station."

or

"This table contains one record for every [transaction / day / customer / product / web page impression / ad impression / download / etc.] and groups/sums [some columns] by [day / week / customer]. Include notes on any logic or filters applied, e.g. 'unknown customers visiting the web site are [excluded / aggregated into the category "anonymous"]' or 'a week is defined as Sunday to Saturday'."

Notes on Business Meaning and Context

This is where you'll define what the dataset shows you and discuss the correct meaning and use of the data. Here, you will also include additional notes to help avoid incorrect assumptions about the dataset.

Example:

"Temperature readings do not take into account humidity, dew point, wind speed or direction, or chill factor; and for this reason, care should be taken when considering the impact of temperature on agriculture in the region, in isolation of humidity and wind speed (for example, frost may occur at zero degrees in low wind situations but may not occur at temperatures well below zero where humidity is low and wind speed high)."

Notes on Known Data Quality Issues

The data dictionary should document any known data quality issues that might affect how the data can or should be used.

Example:

"Records may be missing from the dataset where readings were not taken at a given collection point on a given date. Therefore, average precipitation calculations should use a count of records as a denominator rather than a calculation based on the number of days between two given dates."

Notes on Data Availability

It's here where you document any security and privacy constraints relevant to the dataset and its usage. This equates to documenting the administrative metadata that's relevant to the dataset.

Here is where you define who can access the dataset and for what purposes. You may also define a process for requesting access to the dataset, along with the intended usage. Depending on the sensitivity of the data, this process may require several layers of approval.

It should also be noted that data that is highly sensitive at a detailed level can be masked (certain details hidden) or summarised at a higher level so that the sensitivities don't apply. The data dictionary can point out where the non-sensitive versions of the data can be found if the user isn't allowed to access the sensitive data.

We will look at this in much greater detail in the next section on Data Availability and Access.

Dataset Business and Technical Owners, and Executive Owner

Generally, every dataset has a business owner, a technical owner, and an executive owner. The business owner is usually the person who knows the most about the subject of the dataset and knows the data well. They are usually also the data steward for the dataset. The technical owner is the person who's responsible for the mechanics of keeping the dataset up-to-date and of good quality. This is an IT function, often falling to the lead ETL developer, the data warehouse manager, or the data lake engineer. The executive data owner is usually a managerial level executive (like a VP or GM) who runs the line of business that has the most interest in the dataset.

The data dictionary records this so that others can get in touch with the business owner or the technical owner to ask questions about the dataset. The executive owner won't be involved in any way unless there's an escalation of some issue that requires a decision from them. (APRA calls this person the business sponsor and, under Australian Standard CPG 235, they have particular responsibilities pertaining to data risk and data governance.)

Related Datasets

Ideally, you'll also note where the reader can find datasets related to the one that they're currently requesting.

Example:

"For transaction-level data, see the dataset [dataset name]" or "For daily instead of weekly data, see [dataset name]"

Report Inventory

The Report Inventory is a list of reports and dashboards that are available for access and offers plain language descriptions of what each report contains. You will also make notes about who will find the report most useful and detail the information that the report *does not* provide. As an extra value-add, you could even tell the user where to go to find that missing information, if needed.

You may be able to partially auto-generate this inventory using your BI tool. However, it typically requires manual work to ensure the notes about each report make sense and are searchable. You need to add every report that you generate into this inventory. This means looking beyond the reports your BI tool creates and capturing details for any other processes that produce a report.

Businesses often create this report reactively (after someone realises how much time is wasted looking for reports or maintaining multiple similar reports), which is another example of the "data governance debt" that I spoke about earlier. In fact, almost every organisation I have worked with has said, "We've got hundreds of reports; we don't know which ones are accurate, which ones are important, which ones are even being used!" If you're in this situation, a bit of a "stocktake" might be needed. Some of the Data Catalog tools that I mentioned earlier can automate a big portion of this work.

It's best to set up this inventory early, record information about all current reports, and then add to it as your reporting infrastructure grows.

As I mentioned, each of these three knowledge repositories can be hosted separately, but we're increasingly seeing these three artefacts come together into a "Data Catalog" that is designed to host and publish all of these assets, plus information on technical, business, and administrative metadata and data lineage. A Data Catalog automates the collection of technical metadata by scanning all the databases it can find in your organisation and uses what it finds, and some AI, to make sense of the datasets and have a go at writing for you the first draft of a dataset dictionary for every dataset and data element that it finds. However, invariably, the information that is gleaned or inferred by the tool must be refined and enriched with "tacit knowledge" or "tribal knowledge" that exists in the heads of the people who work with the data.

Component #4 – Data Availability and Access

There are two key components to this aspect of the Data Governance Framework: making it easy for people to access the data they need and making sure people cannot access the data they should not be able to access.

Data Availability is ensured through two key capabilities:

1) Technology that allows people to access and work with the data they need to do their jobs, and
2) Having access to the knowledge of the datasets they are accessing so they can use them correctly.

Data Access is provided through the technologies that your data workers use from day to day: from simple Business Intelligence and database query tools to sophisticated data exploration platforms that allow for many flexible kinds of data visualisation and manipulation.

In general, when we're talking about data access in the context of analytics and Business Intelligence, access to data is always "read-only" — the analyst does not get to change any of the data they are working with, as this breaks the auditability of the data and allows for everything from innocent mistakes to outright fraud. This is also one of the reasons that spreadsheets are considered not best practice for this kind of business reporting and analytics — it's easy to move data, change data, copy data, update data, and lose the audit trail that would allow someone to prove the numbers are correct.

Being able to access the correct knowledge and metadata about the data you're using is just as important as being able to access the data. This is why we've spent quite a bit of time already talking about the importance of things like the dataset dictionary, the business glossary, and the report inventory. Ideally, this information can be accessed with one or two clicks from the visualisation, dashboard or reporting tool itself, so that the correct context is preserved when someone is accessing that information. For example, a report on monthly ARPU should have a link (ideally, the link would be the word ARPU) to the entry in the business glossary for ARPU.

Data Availability and Access also includes making sure that data is not available to people who should not have access to it, for legal or ethical reasons (such as data privacy).

There are three areas to consider when outlining data availability:

1. Legislative Requirements
2. Contractual Obligations
3. Ethical Considerations

Let's look at each of these individually.

Legislative Requirements

In 1890, the US legal scholars Louis Brandeis and Samuel Warren defined information privacy as a human right, with protection under common law. Almost 60 years later, the European Convention on Human Rights also established a general right to privacy and a specific right to information privacy (i.e. protection of personal data) as human rights fundamental to human dignity.

It is worth noting that there are two realms in which your personal data warrants protection. One is the commercial realm: big corporations collecting your location data using apps on your phone or your political preferences based on your social media interactions are constrained by laws set in place by some governments in some jurisdictions as to how they are allowed to use that data. However, when governments collect data on their citizens — such as the Chinese government installing hundreds of thousands of surveillance cameras in major cities connected to facial recognition software, who writes the laws that constrain improper use of that data? Indeed, the advent of "big data" in this sense may be the most significant threat to our freedoms in generations — this is a topic of huge import but one we can delve no further within the scope of this book.

The legal obligation to protect personal data is not a new concept. The challenge here is that legislation is only going to become more complex as we collect more data. In fact, we're already seeing that level of complexity reflected in current laws and regulations. Let's take the EU's GDPR as an example. The following are some of the legal requirements your business has to follow per that piece of legislation:

GDPR Principle	Description
Fairness, Lawfulness, Transparency	Personal data shall be processed lawfully, fairly, and in a transparent manner in relation to the data subject.
Purpose Limitation	Personal data must be collected for specified, explicit, and legitimate purposes, and not processed in a manner that is incompatible with those purposes.
Data Minimisation	Personal data must be adequate, relevant, and limited to what is necessary in relation to the purposes for which they are processed.
Accuracy	Personal data must be accurate, and where necessary, kept up-to-date. Every reasonable step must be taken to ensure that personal data that are inaccurate, having regard to the purpose for which they are processed, are erased or rectified without delay.

Storage Limitation	Data must be kept in a form that permits identification of data subjects (individuals) for no longer than is necessary for the purposes for which the personal data are processed.
Integrity and Confidentiality	Data must be processed in a manner that ensures appropriate security of the personal data, including protection against unauthorised or unlawful processing and against accidental loss, destruction or damage, using appropriate technical or organisational measures.
Accountability	Data Controllers shall be responsible for and be able to demonstrate compliance with [these principles].

What this actually highlights is the importance of your metadata — understanding what datasets you hold and what kind of information they have, especially information that might be sensitive, private, or in other ways subject to legal and ethical constraints. Moreover, when new legislation comes into force, if you know what kind of datasets you hold, you can easily identify datasets that might be impacted by the new legal framework.

Any restrictions you put in place must be enforced in a permissions model inside a technology that stops people from accessing data they should not be able to see. It's not enough to have a warning sign; the road has to be blocked off. This is done by defining permissions in the database itself or in the tools that people use to access the data (see Data Security). There is also an emerging class of software that provides an access layer over multiple data sources — data virtualisation — so that you only have to define the permissions once, and then apply them to all data sources. If there's ever a data breach that is investigated by authorities, you've got to be able to show the principles that were used to define who can see what data and then show how those principles were enforced in terms of permissions and data security.

Data Access also has to be auditable. This means you need a way to keep track of who has accessed what data and when. Database logs usually store this information, but these logs are also often streamed to a central repository (Splunk is a technology that's often used for this) where it's easy to run queries and reports across all data source logs to construct a picture of what data has been accessed by whom. Again, in the case of a breach, regulatory authorities will expect you to be able to produce these audit logs.

Contractual Obligations

It's becoming more common for businesses to access and use data from other organisations. Sometimes, this is done under a commercial arrangement, like a subscription to a dataset on, say, frequent fliers for a sister loyalty program or a credit bureau providing your credit score to a bank. In other cases, it may be open source data such as from the Australian Bureau of Statistics.

Access to and use of these datasets usually comes with restrictions on their usage, as part of the contract under which the data is accessed or obtained. In the case of the credit bureau, for example, the data might be restricted to a single use and not allowed to be kept or to be used for a different credit application from the same person later on.

As an example, NBN Co is among the many businesses that I've worked with during my career. During my time with them, we purchased a dataset with information on every business in Australia. Our contract with the vendor specifically outlined how we were permitted to use that data. For example, we could use it for direct marketing, but we could not share it with external marketing companies unless there was an additional "side letter" providing that permission and limiting how the marketing house could use the data.

Many data providers will apply similar restrictions in their contracts. Your data governance policy needs to account for these restrictions, and you must enforce relevant database permissions relating to those datasets. Again, this allows you to demonstrate due care and diligence should an issue arise.

Breaking your contractual obligations doesn't just mean the potential end of your relationship with a vendor; it is likely there would be legal action to recover costs and damages if the data was used incorrectly. It can also result in reputational damage, both in the eyes of other vendors and the public.

Ethical Considerations

If we revisit the Cambridge Analytica scandal, we see a perfect example of a company that failed to take ethical considerations into account, even though they adhered to the letter of the law. Two years of investigation found that the company committed no legal breaches in its use of data. However, the ethical breaches they committed fuelled the scandal to the point where the company had to close its doors.

Its former COO, Julian Wheatland, sums it up best when he says:

"Cambridge Analytica made many mistakes on the path to notoriety. But its biggest mistake was believing that complying with government regulations was enough and thereby ignoring broader questions of data ethics and public perception."

Ethical considerations apply most frequently in the contexts of data sharing and data commercialisation. If you breach these considerations, you run the risk of souring public perception even if you didn't break the law. Upon reflection after his Cambridge Analytica experience, Wheatland came up with four key ethical principles to follow to ensure high ethical standards as well as the public perception of high ethical standards, which is almost as important:

1. **Transparency** – Clearly state what information you hold on your customers, how you use it, what you use it for, and why. Create a set of guiding ethical principles and make them public. This improves public confidence in your use of data, in addition to giving you a set of guidelines to follow.
2. **Effective Management** – Set up internal procedures, with full support from the board, that guide employees. These will outline

how they should use new tech and data, in addition to providing them with a set of ethical principles to follow. Whenever somebody highlights a lack of clarity in these processes, elevate it to the Data Governance Steering Committee (or a similar body) and re-clarify before signing off on the procedure. Having procedures in place means little if there's no management and accountability behind them.

3. **Honesty** – Consider appointing a Chief Ethics Officer. Their role is to ensure that the company is adhering in practice to its own ethical principles. This person keeps the organisation honest in all of its work with data. But as importantly, they must anticipate potential data issues and help the organisation adapt to them.
4. **Choice** – Any individual for whom you hold data must have the ability to easily opt out. Wheatland points out that this is one of the things that Cambridge Analytica didn't do. Despite this being okay from a legal standpoint, that lack of choice came back to bite them when the scandal hit. This choice should extend to third party companies that may acquire your data and also enable an individual to rescind and undo any previous agreement to have their data shared.

Perhaps the key point to make here is that meeting your legal obligations does not automatically mean that you're meeting your ethical ones.

Component #5 – Data Security

The Data Governance function may be charged with developing the organisation's Data Security Policy. As with Data Access, your Data Security Policy needs to take into account — and make reference to — relevant regulatory requirements, contractual obligations, business requirements, and ethical considerations.

Your data security policy and its implementation are likely to be required as part of evidence should you find yourself facing legal proceedings. The lack of a Data Security Policy may lead to the conclusion that your organisation was not serious about complying with relevant legal or ethical standards. This can lead to an increase in penalties applied should a legal or ethical breach occur. In many cases, the organisations that regulate your industry will also outline their security requirements, which you can build into your policy.

In Australia, APRA outlines its expectations of any APRA-regulated organisation relating to data security through Prudential Standard CPS 234. For example, when discussing the organisation's data security responsibilities, it stipulates:

"An APRA-regulated entity must clearly define the information security-related roles and responsibilities of the Board, senior management, governing bodies and individuals with responsibility for... information security functions."

APRA's guidelines are actually quite pragmatic and are designed to take into account the size and complexity of an organisation, the potential value of their data assets, and the potential harm from a data breach. For example,

"An APRA-regulated entity must maintain an information security capability commensurate with the size and extent of threats to its information assets, and which enables the continued sound operation of the entity."

One of the best ways to do this is to break down your data into sensitivity levels and then manage each category of data with the appropriate security framework and processes. Typical of these sensitivity levels are along the lines of the following:

- **Public** – This covers any data or information that's already publicly available. Product brochures and the content on your website fall into this category.
- **Private** – Any information that isn't for circulation outside of the organisation but where the disclosure of the information would be more of an inconvenience than a legal or ethical breach. Examples of private information include internal memos and project reports.
- **Confidential** – This is information for which access is limited to a subset of employees and third parties. Disclosure of this information could lead to direct financial loss and the potential for penalties. Examples include personally identifiable information

(PII), staff payroll details, credit card numbers, details of customer complaints, or employee performance reviews.

- **Highly Confidential** – This is information that should only ever be disclosed to specially authorised personnel with a "need to know". If the information goes beyond these personnel, it will create significant issues for the organisation. This could include heavy financial penalties, in addition to the restrictions placed on its activities. Examples include encryption keys, health records, and details of corporate acquisitions, mergers, or legal proceedings.

Given the different levels of sensitivity that apply to different kinds of data, it makes sense not to apply a single data security standard to all of the data that flows through your organisation. As part of your data governance policy, it is advisable to place data into one of these categories and have appropriate security measures in place for each.

Enforcing Data Security

Ultimately, data security is how you enforce your rules on data availability. We can divide this into four components:

- Authentication – How do I know that you are who you say you are?
- Authorisation – Now that I know who you are, what are you allowed to see?
- Access Security – How do I prevent *unauthenticated* or *unauthorised* parties from accessing data they should not?
- Auditing – Who accessed what data, and when?

Authentication covers the methods used to provide access to the data. Typically, this is a username and password to log into a system. As more data and applications move to the cloud, we see increasing use of cloud-based authentication services that allow "single sign-on" authentication across multiple platforms which check the authentication service before giving you access. Sometimes called "authentication as a service" or AaaS, examples include Okta, OneLogin, Auth0, and AWS Cognito.

We are also seeing the increasing use of two-factor authentication or 2FA. 2FA means that more than one element is required to confirm your identity. In a building, this could mean swiping your security card and then also entering a PIN. For technology services, it usually means providing a login and password and then providing an additional step such as entering a code sent to your phone.

Authorisation means ensuring that, once a person (or application, API call, external data connection, etc.) has authenticated, they can only gain access to the data that they should be allowed to. This might be enforced in the database itself using roles and permissions, or it may be enforced by a common security and permissions layer over all your data assets.

Access security refers to the enforcement methods to prevent unauthorised access to your data assets. It covers the physical security controls you have in place for your data centres and backup data. If somebody can walk into a data centre and walk out with a hard drive with nobody stopping them, this suggests poor physical security measures. It also covers the methods that you use to protect online data, such as firewalls, appropriate passwords, and any cybersecurity measures you take such as encrypting data. Encryption, in particular, becomes very important when data is in transit (travelling over a network which could be packet-sniffed).

Finally, auditing means tracking and being able to report on who accessed any specific item of data and when. Often, this is critical to detect malfeasance such as internal fraud. I can recall an example of where workers in a particular government department were accessing details of inactive registered businesses and then using those details to funnel money into their syndicate. What they didn't know was that every time they accessed a business record, it was being tracked and before long, they were exposed and prosecuted.

Auditing of data access can take on a life-and-death significance in some interesting cases such as witness protection. If a record in a police database is accessed by an unauthorised person, an alert must be immediately created and acted upon — as the details may have been accessed in order to provide them to the criminals involved, who intend to track down and harm the witness who is scheduled to testify against them.

Component #6 – Data Risk

One of the essential aspects of managing data is understanding and managing data risk. Data risk should be considered across a number of forms (some of which we have alluded to already).

Regulatory and Compliance Risk

When we look at the Data Governance Roadmap later in this book, one of the key steps is an Environment Scan. This means taking the time and effort to be informed of any legal requirements, ethical considerations, and community expectations that impact the way you use the data you have.

Of particular relevance in Australia are the expectations captured in APRA legislation called CPG235. While this legally applies only to APRA registered organisations such as banks, insurers and superannuation funds, the guidelines are instructive to any organisation and are worth reviewing (in detail) and using to help inform your approach to data risk management, regardless of what vertical you're in.

Other examples of regulatory requirements might include the need to retain certain data for five or seven years in case a taxation authority requires access to it as part of a review or investigation.

Personally Identifiable Information (PII) Risk

One of the key areas to be aware of is the regulatory requirements regarding the treatment of Personally Identifiable Information (PII). At the moment, we are seeing various jurisdictions formulate their own set of PII requirements. Often, these are legally applicable to data you may hold on a country's or region's citizens, even when you are not doing business in that region. For example, the European GDPR data privacy requirements apply to and protect the data of all European citizens anywhere in the world, regardless of where the data is captured, stored, or used. Penalties for lapses are significant, up to €20 million or 4% of the company's global annual turnover, even for companies not based in Europe.

Similarly, The California Consumer Privacy Act (CCPA), the UK's Data Protection Act, and Brazil's General Data Protection Law ("Lei Geral de

Proteção de Dados" or LGPD) apply to their citizens' data regardless of where you're doing business.

This can all make for quite a complicated regulatory landscape, but one you can ill afford to ignore. I believe that, in time, many of these standards will merge into a commonly accepted global standard for data privacy. We may even see the emergence of blocks of countries that cooperate on data privacy and then share privileges which are denied to countries who do not sign up to the same level of transparency.

In the meantime, however, it is incumbent upon organisations to ensure they are compliant and are seen to be making efforts to be compliant with relevant PII requirements.

Risk of Data Breach

We've referred to aspects of data security in the section immediately above; however, another aspect to managing data risk is ensuring that you have defined processes in place to respond to various levels of data security compromise.

Data security compromise can take a number of forms, such as a disgruntled employee sharing sensitive information with the intention of doing harm to their employer or the employer's reputation; phishing and spear-phishing attacks that seek to get someone with access to sensitive data to enter their credentials into a bogus site which then uses those credentials to access that sensitive data; penetration and exfiltration of data via "traditional" hacking; physical compromise through the loss of a storage device like a thumb drive or laptop; all the way up to methods that might be more associated with state actors, such as blackmail or extortion.

It should be noted, however, that most data breaches (70% in fact) are due to accidental compromise rather than malicious intent. Furthermore, your data could be exposed by your business partner organisations rather than by your own mistakes or negligence. This risk is becoming exponentially more significant as data sharing becomes more common. Currently, the only controls in place are an "agreement" or checklist that says all the companies you share data with will be responsible for it — but their cybersecurity regime may not be as mature as yours, and your data gets compromised from their negligence or lack of knowledge. Some companies (an example is Paris-based cybersecurity firm CybelAngel) can actually scan for your exposed data in multiple perimeters (open web, dark web, exposed servers, and databases, etc.) to detect data leakage that might be via a third party or a data partner outside of your control.

As noted above, a key aspect of managing data risk is ensuring that you have processes in place to respond when something goes wrong. You'll recall that in the email example I shared earlier, the biggest part of the cost of rectification was that there was no process in place to respond to a data breach. Part of the role of data risk management is to consider the various scenarios in which things could go wrong and have defined responses for each.

Risks Due to Poor Data Quality

Risks arising from poor data quality include the likes of:

- the risk of reporting incorrectly or late to regulators
- the risk of poor decisions based on inaccurate or late-to-arrive data
- the risk of errors in estimates of key business metrics or financial ratios

Risk of Data Loss or Corruption

Data loss or corruption can happen when physical hardware such as disks or servers are damaged or "go down". It can also happen when someone with write or update permissions to the data accidentally runs a command that deletes or corrupts records unintentionally.

In either case, standard procedures dictate that business-critical data is backed up regularly and that processes exist to restore the data to a given point in time before the corruption occurs. Sophisticated tools are available to help with this, and an entire industry and set of best practices has emerged around this, including "high availability" to ensure that a data source is replicated in real time across multiple locations so that one is always available should the other be affected by an outage or other incident.

With the migration of much data to cloud services such as GCS, Azure and AWS, which are offered with backup and high availability "built-in", there is still a need to ensure that recovery is possible in the event of data corruption — often, this is achieved through snapshotting data at points in time and having data replicated across multiple zones.

Risk of De-anonymisation

When sharing datasets with other organisations or even publicly, any PII should be masked or hashed to avoid allowing individuals in the dataset to be identified. However, there is a danger that someone can combine a few different anonymised datasets and, by linking common attributes, derive enough information to de-anonymise the data and identify individuals, their PII, and other information about them.

In Australia, a government department released anonymised health records to assist researchers analysing health trends across the population. Some university students obtained several similar datasets and, by combining these, were able to match enough fields to identify individuals. Fortunately, their work was not used for any nefarious purpose (such as selling PII on the dark web), but it does highlight that care must be taken to ensure that anonymisation is undertaken rigorously and, perhaps, with the assistance of experts in the field.

Data Remanence

This refers to data that was supposed to be deleted but copies still remain. This might create a risk of legal breach (the subscription has expired, so any data still kept, even accidentally, is a breach of contract) or of accidental discovery and exposure of the data at some later time.

Data Sovereignty

Data sovereignty refers to the fact that data is imposed upon by the laws and practices of the nation in which it is persisted (physically stored). Many countries require certain kinds of data to be only stored within the country and not sent outside its borders. This is particularly true of sensitive information that might be of interest to a foreign government, whether it be corporate financial records or relating to defence and intelligence services.

In other cases, companies or organisations may have policies that restrict data from being moved outside the country.

As part of the Environment Scan mentioned above, you need to check whether the government of the country or countries in which your data is captured has requirements around where the data can be stored. Even if there are no hard legal requirements, ethical considerations can also come into play. For example, it is fair to assume that any data that is sent to or through China or the USA in an unencrypted format can be scanned by their respective governments (intelligence agencies) at will. A university I worked with refused to allow any student data to go outside of Australia in case a foreign government scanned the data and used it to form a profile (say, for criticising a government, a head of state, or a religion) which could have led to an arrest if the student ever travelled to that particular country.

Data Risk Across the Data Lifecycle

As discussed in CPG235, each of the above data risks (and any that apply to your organisation which I may not have identified) need to be considered at each stage of its life cycle, ensuring that appropriate controls are implemented.

The data life cycle can be considered as the Capture, Processing, Retention, Publication, and Disposal of any data.

Component #7 – Master Data Management

Master Data Management, or MDM for short, refers to managing a single trusted version of the attributes and hierarchies for key business entities. You can think of an MDM system as a hub that receives all data updates and applies the appropriate rules to check for validity and trustworthiness. This hub then distributes that verified information to all relevant systems and business processes. A good example might be customer address and contact details. This information could be updated in any number of places — the call centre, a web form, outbound marketing, incoming email, and so on. Regardless of where the update comes from, it needs to be recorded into a single source of truth, and then every system and process that requires those details needs to be able to access the updated version of the truth.

We sometimes refer to this as the Golden Record. The benefits of this approach include the following:

- All areas of the organisation work from the same information, with the MDM system mastering and sharing the most trusted and up-to-date version of the data.
- It improves data quality.
- There's less risk of different versions of the same information populating different systems. For example, an MDM ensures that any updates a call centre operator makes get fed into relevant marketing and sales databases.
- An MDM reduces point-to-point integration activities between systems. For example, a CRM does not have to push its updates to all other systems because the MDM centralises the process. Over time, this can provide significant savings in IT spend on system integration.

Generally, Master Data Management (MDM) is implemented in an organisation using specific software designed for this purpose. An MDM is a specialised system that collects reference data from an array of other systems. The MDM will then determine the most trusted version of the data it collects. Typically, this is the most recent version of a particular attribute, but different sources can be evaluated for the level of trust (for example, the information provided in a web form will be less trusted than the same information provided after the person has logged in and confirmed their identity). Where there is a conflict, the MDM consults its "survivorship rules" to determine which version of the information should be trusted and mastered. It then stores this chosen attribute and publishes it to all relevant systems across the organisation (or makes it available as an API call).

Component #8 – Data Quality

The Data Management Body of Knowledge (DMBOK) defines data quality as follows:

"Data is of high quality to the degree that it meets the expectations and needs of data consumers. That is, if the data is fit for the purposes for which they [intend to use it]."

"Fit for purpose" is the key term in this definition. Fit for purpose can mean different things depending on the purpose. For the finance department, it may mean that the figures are 100% accurate, auditable, signed off by the CFO, and suitable to go to a regulator. For the marketing department, a customer segmentation model with 85% accuracy might be "fit for purpose."

Furthermore, because the same data may be fit for one purpose but not for another, the Knowledge Management assets that we discussed earlier become even more important because this is where you capture and make available the comments on data quality that data workers need to know about when they want to use the data.

We can evaluate and measure data quality across seven dimensions:

1. Accuracy – Does a piece of information reflect reality?
2. Completeness – Does it fulfil your requirements for what's comprehensive (what needs to be known)?
3. Consistency/Integrity – Does information stored in one place match related data stored elsewhere?
4. Timeliness – Are the records updated and accessible in a timely manner?
5. Validity – Do the values follow the applicable business rules (e.g. format, range, legitimacy)?
6. Uniqueness – Are there invalid duplicate values?
7. Environment – Have there been changes to the circumstances under which the data is collected, which affects how it can be used?

Let's briefly examine each of them.

Dimension #1 – Accuracy

The Accuracy dimension evaluates how closely the piece of information reflects reality.

As an example, I was working once with a dataset on Australian businesses. One of the records was for BHP, and one of the attributes was their primary business vertical. This dataset had assigned BHP to "Oil and gas exploration". Now, BHP does undertake this activity, but it's not their primary business which is mineral and ore extraction. The data was not accurate, and this affected the "purpose" for which we wanted to use the data, which was targeting companies in specific verticals.

Simpler examples include recording the wrong address or surname for a customer. At the more extreme end is the example of the sensors on the Boeing 737-MAX. They were prone to reporting a stall event when the aircraft was not stalling — inaccurate data. Unfortunately in this case, the objective source of truth (the pilots) was overridden by the way the system was designed, and this led tragically to two aircraft crashing and many deaths. Data accuracy can really be a life-and-death issue in some situations.

The key challenge here lies in determining the source of truth that you compare the data to. In the case of information about a customer, it's the customer themselves who's likely the source of truth. If you've ever had a call centre operator ask you to confirm your address, this is an example of an organisation confirming the accuracy of its data against the source of truth.

Incidentally, this is also an example of a proactive method of determining accuracy. It takes time and money but typically results in better data quality. The alternative is the reactive approach, where you fix a data quality issue once it's discovered. This is cheaper because you don't have to spend time and effort auditing datasets for accuracy against an authoritative source or objective truth. Clever organisations will build mechanisms to check the quality of data into their business processes so that this "reactive" approach yields greater results.

On the other hand, a customer may have an incentive to provide false information such as about their credit history. In this case, the source of truth might be a credit reporting agency that can provide accurate data on the customer's credit history and rating.

Dimension #2 – Completeness

Data Completeness is a measure of how many elements of data are missing where they might be expected. This could be missing records (e.g. a POS transaction was not passed to a financial system because there was a glitch at the terminal), or missing data within records (e.g. we have the person's email address but no phone number).

The concept of "fit for purpose" is critical in evaluating whether data completeness matters for a given use of the data. For example, you may not have a record for the middle name of a customer. However, this isn't likely to matter in marketing communications, where you'll likely only use their first name and surname. In this case, the data is "100% complete" for the purpose of a mail-out.

The challenge here arises when you have multiple departments that need to use the data. A missing middle name may not matter for basic marketing purposes. However, it may matter if you need to send out more official documentation, such as a health record.

Again, the importance of our knowledge management assets — the data dictionary in this case — is highlighted. It would be crazy for every person who wants to use a dataset to have to check it for completeness by running queries and analysing the results. The first person who runs some data completeness investigations on a dataset must record their findings in the data dictionary so that those who come after them can benefit from that knowledge and save huge amounts of time.

Dimension #3 – Consistency

Data consistency is high where all systems have the same data on the same entities. This is also sometimes called Data Integrity. There is a lack of data integrity in system B if system A has updated records while system B does not (where system B requires the updated records for the data to be fit for purpose). To put it another way, there's a lack of data consistency between system A and system B.

The use of MDM proves its value here as it is designed to keep all of the subscribing systems up-to-date with any changes that are made in a system of record (and validated by the MDM). The alternative is that if an update occurs in System A that isn't reflected in System B, you have an integrity issue; and those who use System B aren't working from up-to-date data.

The classic reporting issue of different reports showing different results when they are supposed to be showing the same thing is an example of a problem with data consistency.

Dimension #4 – Timeliness

Is the data up-to-date and available in a timeframe that ensures that it's fit for purpose?

There are two components to consider when it comes to timeliness:

- Is the data updated from the source of truth in an appropriate amount of time?
- Does that data become available to the user in a timely manner?

Fit for purpose is a key consideration here. For example, live data on share prices or future markets may only be useful for a few seconds as it can change so rapidly, and the opportunity to make a decision (to buy or sell at that price) is gone. In fact, high-speed traders (using powerful computers and algorithms to automatically buy and sell on the market) actually pay a premium to rent data centre space as close to the exchange as possible, or even in the same building! Even though data over fibre optics travels at the speed of light, that few nanoseconds mean the difference between getting the deal or missing out to someone else.

On the other hand, once trades are made, the data has value in a different way — it allows pricing trends to be spotted, as well as anomaly detection which regulatory authorities use to detect possible fraud and insider trading. This historical data actually never goes out of date.

Most reports source data from a data warehouse that is updated overnight, so the data is always at least one day old. Usually, this is not a problem, and many reports only look at monthly summaries of business activity anyway.

Dimension #5 – Validity

This is a check for whether the data is consistent with what is expected or required of the data for it to be real. For example, a postcode in Australia will always contain four digits and no letters. Any other kind of entry is not valid. A date with a month larger than 12 or a day higher than 31 would also be invalid.

Data validity checks are an easy way to ascertain if there are data quality issues in a dataset.

An important note here is that data can be valid without being accurate. For example, a person's name that has been misspelt will be valid (we're expecting text and we have text) but not accurate.

Dimension #6 – Uniqueness

Uniqueness is violated if there are duplicate entries where there should not be duplicate entries.

The most common example is if an organisation holds multiple customer records in a system for a single customer. This can be a minor issue — for example, someone receiving more than one letter or email about the latest offers — or a major one. During the 1997 Asian Financial Crisis, some banks in the region discovered that their customers had taken out loans against the same asset (like their house) with more than one bank. They invested the funds and, when the downturn hit and their investments went bad, they defaulted on their loans. The banks then discovered that several other banks were trying to claim the same house to recover their funds. In the country I'm thinking of, I worked on a project for the Central Bank to ensure that this could not happen again. In effect, the Central Bank implemented an MDM into which all the banks had to match their customer data and also any assets that were used as collateral for a loan.

Dimension #7 – Environmental Change

Environmental factors can also come to bear on how data can be used or interpreted to support decisions or measure performance. In 2004, I was consulting with an airline on automating their 100+ page monthly board report. The SARS downturn in the airline industry in 2003 meant that their "year-on-year" figures looked spectacular in 2004! We had to implement a workaround so that 2004 year-on-year figures were calculated against 2002 data to provide a more realistic view of the airline's performance in 2004. The 2003 data was affected by an external or environmental issue, which meant that it could not be used in certain ways (it was not fit for those purposes). Ideally, this information is captured in the business metadata repository, such as the dataset dictionary, so that everyone accessing the data is made aware of the anomaly.

We've seen other examples of environmental change affecting data in the discussion around global warming. As cities grow larger, a temperature reading station that was once in the middle of a farm is progressively surrounded by more and more asphalt and buildings, and the temperature readings increase as a result. If a person interpreting the data is not aware of the impact of urbanisation on the temperature readings, they could easily come to a wrong conclusion. That is, if the purpose of using the data is to figure out what to wear today, the data is fit for purpose. If the purpose is longitudinal analysis over many years to identify changes in global temperatures, the impact of urbanisation on the temperature readings has to be taken into account and the readings adjusted for that impact before the readings can be used for that purpose.

Benchmarking the Dimensions

In order to monitor the level of data quality in key datasets over time, we can define benchmarks for each of these data quality dimensions. The benchmark is the "initial reading" of quality in a particular dimension; the same tests can be run at regular intervals to determine whether data quality is going up or down. This, in turn, helps to inform the people using the data whether or not it might be fit for a given purpose. Typically, these dimensions will look something like this:

- **Accuracy** – what percentage of incorrect values makes the data unfit for a particular purpose (of course, this can vary depending on the purpose for which it has to be fit)?
- **Completeness** – what percentage of missing values can you accept before the data is unfit for your purpose?
- **Consistency** – what level of inconsistency between systems is okay?
- **Timeliness** – how late can the data be updated and available and still be fit for a given purpose?

- **Validity** – what is the impact of data that is outside of the valid ranges?
- **Uniqueness** – what impact do invalid duplicate values have?
- **Environment** – do users need to be aware of any changes in the environment that affect how the data can be used?

Perfection in data quality is almost never achieved and so, this should not be set as a goal. Instead, we begin by ascertaining the quality of our data in its current state against these dimensions and then determine where the biggest "bang for the buck" is going to be. Then, we get to work on addressing the data quality shortfalls.

The Seven Steps for a Data Quality Uplift Project

Knowing the state of your data assets — their trustworthiness, which is based on the quality of the data — is essential for being able to confidently make data-based decisions and is foundational to successful digital transformation, process automation, and value from data.

In my observation, there are often times when an organisation needs to undertake a project specifically to address a data quality issue. I referred earlier to the "pendulum effect" where organisations tend to neglect data governance and data quality in order to reduce short-term effort, overhead and cost until a crisis occurs and they are forced to address the accumulation of data issues — at which time they run out and recruit a team or engage an expensive consultancy to jump in and fix the issues. If the project or engagement is well executed, there will be a sustainable set of best practices in place that can help the organisation to maintain its data governance and data quality going forward. But eventually, someone unfamiliar with the history decides there are ways to cut corners and cut costs and goes about reducing the "overhead" of good data governance — and so, the cycle repeats.

Not every dataset needs to be of perfect quality. Indeed, the pursuit of data quality is, in most cases, a never-ending quest, so you have to target your efforts at the areas that are going to have the greatest beneficial business impact, the "best bang for your buck." At some point, the law of diminishing marginal returns applies to data quality also, and at this point, you would dial back on projects to address data quality and simply maintain business as usual.

A data quality project will require investments of both time and money. However, it's also the perfect chance to put the adage of "never let a good crisis go to waste" into practice. There are a number of approaches to data quality projects, and you may wish to research this in some detail if you're going to invest serious time and money into such a project. However, the following steps provide an overview of the essential components for a successful Data Quality Uplift project. These can be used to help determine how your team will define and approach a data quality project or can be used to "sanity check" a proposal from an external party.

Step #1 – Determine the Scope

Which datasets fall into the scope of the programme? And as importantly, which of the seven dimensions of data quality are to be addressed by the programme?

The goal here is to take a dataset, or a number of closely related datasets, from a state where they *are not* fit for a given purpose to a state where they *are* fit for purpose. The fitness for purpose test will tell you which of the dimensions of data quality you're trying to address. It could be that customer phone numbers and addresses are not accurate — they do not reflect reality — and this is impacting your ability to market, thus, impacting the bottom line. Or it may be that data between different systems is not consistent and this is surfacing conflicting reports to senior executives.

One key to success here is that you only include closely related datasets into a single uplift programme. The inclusion of unrelated sets creates complexity, so you should split these into different workstreams or even into a programme of their own.

Step #2 – Determine the Current State, Target State, and the Gap Between the Two

Once you know which data dimensions fall into your scope, you need to determine where they currently stand and where you need them to be. In other words, you set a target state in terms of what will make the data fit for purpose. From here, you determine the gap between this target state and the current state. As part of this process, you need to define the answers to these questions:

- How will we actually fix the data?
- How will we measure progress along the way?
- How will we know when we have arrived?

Step #3 – Identify and Measure the Benefits

What are the hard business benefits and outcomes that you stand to gain from this data quality uplift programme? Outline the benefits, outcomes, and impact of the work that will take place for each dimension that's in scope. You can then use this to prioritise tasks in the programme based on both their complexity and the value that they bring to the organisation.

A simple matrix such as the one below is a good way to determine the priority of your efforts and also to identify that point of diminishing returns.

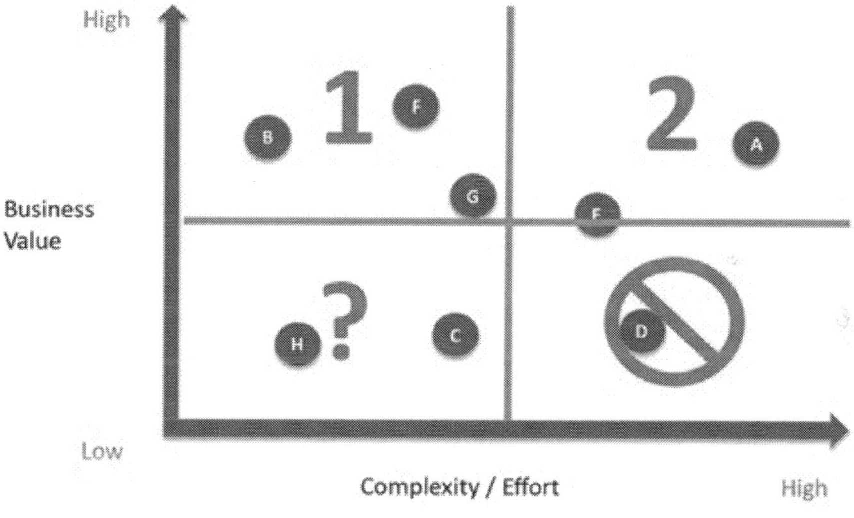

Step #4 – Identify the Root Causes of the Data Quality Issue

This requires you to go upstream so you can identify the ultimate source or sources of the data you want to fix. You also need to understand the data lineage — any changes that occur to the data before it gets to the repository (e.g. data warehouse or data lake) where it is accessed for the purpose defined in the scope.

Once you've identified the upstream root cause of the data quality issue (the "source of contamination"), there are several steps for you to take.

Engage the owner of the process as well as any stakeholders affected by the data quality issue.

Determine if it's viable and cost-effective to fix the root cause and, if it is, determine what you need to do to implement the fix.

In some cases, fixing the root cause may require the creation of a separate project that focuses specifically on this piece of work. Note that this is often done by a different team (e.g. application development, architecture, ETL/EDW, etc.) and that project should be owned and managed outside the Data Quality Uplift workstream (but may be financed out of the bucket that appeared after the crisis).

You may also need to invest time into achieving buy-in at the executive level across a number of business units, articulating why this root cause is such a large issue and the benefits of fixing it. Seek out an advocate in the affected lines of business, in IT or in the PMO, who can see the bigger picture and help agitate for the changes that are needed to get that separate project over the line.

Step #5 – Break It Down and Execute

As part of the previous step, you started to identify the steps needed to fix the data quality issue. It's here where you solidify those steps and assign ownership of them. This may involve engaging your Project Management Office to get their help with managing the project.

The specific steps will vary depending on the nature and scope of the issue. You may have some steps that you can automate, where others require manual action. Some steps may involve one-off fixes, with others requiring the creation of a process that is triggered only under certain circumstances. It's also here where you'll identify and purchase any tools or technology that can help you to fix the problem. You also need to decide whether you are just fixing the issue so that data going forward is of higher quality, or whether you also need to go back and fix the historical data that has already been collected (and if so, how far back in the historical data you decide to fix).

Step #6 – Define the Lessons Learned

As the team executes these steps, they are bound to come across opportunities to learn and apply new ideas to other areas of the business. Preserve this knowledge so that it can help the next team that works on this dataset and look for colleagues with similar problems who can benefit from your experiences.

You can break these down into project learnings and data learnings.

Project Learnings: the PMO should have a project-oriented Knowledge Management (KM) methodology to retain what project managers learn that is of value to other project managers.

Data Learnings: Ensure the Dataset Dictionary and any metadata repositories are updated with what changed, why, how, and when. (Note in detail especially if data was fixed back to a certain point in time as this impacts data consistency — any future longitudinal analysis or time series analysis of the dataset must be aware of how historical data was fixed to a certain point.)

Step #7 – Communicate the Successes and Benefits

This is where you validate the expected benefits that were outlined in step 3. Compare the actual results against what was expected to see which outcomes were realised or even exceeded. If you do exceed an expectation, consider this a point of learning that you can apply to your next uplift project.

Finally, communicate and celebrate the benefits achieved through the programme. Not only does this benefit your team, but it also gives you political capital for future data projects.

Data Quality Guerrilla Warfare

An alternative or complementary approach to the "big bang" Data Quality Uplift Programme is what I call Data Quality Guerrilla Warfare. This means looking for small, tactical wins for data quality, adjusting a process that is allowing poor-quality data to be captured or distributed, and one-off fixes applied to a given dataset that will result in sustained improved data quality.

This approach often requires ad hoc, tactical collaboration across teams without a formal programme or project (and hence without funding). People will often have to provide assistance at their discretion (it may not be part of their formal job description and they can say no, so you've got to gain buy-in using friendship, persuasion, and mutual interest). Gain buy-in by "winning hearts and minds" to the overall benefit to the organisation. Sell the vision!

This approach still requires an assessment of the current state, the target state, and the business value/impact of completing the project; it can also provide 'political capital' for future and larger data quality initiatives. And again, advertise and celebrate the small victories!

In the next chapter, we'll briefly examine some of the technology you may need to help underpin your Data Governance Framework and initiatives.

CHAPTER 10

Data Governance Technology

Before considering technology investments related to data governance, it is critical to understand that buying and implementing new hardware and software will not solve data governance, business processes, or organisational/cultural issues. Too often, we tend to see technology as the solution to a problem rather than as a tool that we use to implement the solution! Data governance is not a technology challenge. However, good technology can make it easier to implement and optimise your data governance processes.

If you've worked in IT for a while, you've probably seen the typical process that often triggers a new technology investment. There's an accumulation of issues that people discuss in various forums, then an IT project pops up on the radar screen and everyone starts to hear about it and think "that will solve this problem." So they delay actually fixing the problem, waiting for the technology to come along and expecting that the new technology will solve the problem. It never does because the actual problem is with the organisation, the culture, or the way of working.

Your organisation's technology issues, to the extent that they are real, are actually the easiest issues to solve. It's resolving the cultural and organisational challenges revolving around data that are almost always the key problem. This involves changing the way "we've always done things," dealing with the inertia that comes from that and potentially also confronting political agendas within the organisation.

There are entire methodologies to assist with good technology selection — such as documenting functional and non-functional requirements and target state business processes to be supported by the technology. These are valuable approaches, and if there aren't people in your organisation who can run with this in an experienced and disciplined way, it's best to get outside help from a consultant or two who can advise (Trust me, this is much cheaper than selecting the wrong technology and investing all the time and effort trying to make it successful.).

You also need to map out whose roles will be impacted by the change, and create a change management plan to ensure a smooth transition for them, including what training needs to take place. Also, outline the benefits you expect the technology to bring in the context of your data governance efforts. As you roll the technology out, communicate regularly with stakeholders and ensure they have everything they need to use it properly. On a technology rollout that is complex or affects a large number of users, engaging a professional Change Manager, even if on contract, can have a huge impact on the success of the rollout.

Aspects of Data Governance (and the Technologies That Fit Them)

There are a number of specific data governance related processes that should be supported by appropriate technologies or a technology ecosystem.

These are noted and explained below:

- Technical metadata management – this means an app that can "spider" out across the organisation and find every database, query it for its schemas, tables and columns, and ideally provide some intelligence around this by identifying sensitive data like credit card numbers, PII and so on. Data catalog applications are usually good at doing this.
- Data Lineage – where data is copied and transformed between databases, the tool(s) that are used to do this should be able to be interrogated to return the end-to-end lineage of the data. Ideally, this extends into any BI tools and ultimately into the reports themselves, so you can trace any item in a report back to its operational source system. Again, a good data catalog application should be able to do much of this automatically by collecting the technical metadata from the tools that are transforming the data. If this can't be automated, it's a massive undertaking and you'd be better off selecting only the most critical data to document the lineage manually.
- Business Metadata Management – this refers to the Business Glossary and the Dataset Dictionary. Typically, these would be instantiated using a Data Catalog, and the Data Catalog would

usually also be the tool doing the technical metadata and data lineage collation.

- Report Inventory – often, this is part of the Data Catalog. To the maximum extent possible, the collation of reports and the information about reports (how often they are used, who uses them, and so on) should be automated. It's just not sustainable to try and keep track of this manually.
- Data Quality tools – specific apps to measure data quality (as discussed above) and optionally to apply fixes to data quality in the analytical repositories (ideally, the data should be fixed in the source, but that's not always possible). These tools should also allow you to track progress in improving data quality over time. Data Quality can be managed and tracked in a standalone app or as part of the Data Catalog suite.
- Data security and access – this must either be enforced in the database permissions using roles and users, or it can be done in a layer above the databases that sit in a different technology. A number of options are available from traditional vendors, new startups like Okera, and cloud providers.
- MDM – again, this can be mastered in a standalone app or as part of the Data Catalog suite.
- Data access auditing and alerting – using the database logs to track who accessed what data and when. Tools like Splunk are good for this but not always cheap. If sensitive data access needs to generate real-time alerts (such as the police protection example mentioned earlier) this should be factored in as a requirement.
- Data Issue tracking and resolution – a mechanism for people to create workflow, track and close support tickets relating to data issues such as poor-quality data, incorrect or unavailable data, and so on.
- Data Risk tracking, escalation, and management – a mechanism for people to create, workflow, track and close tickets relating to data risk that they might identify either in their day job or as part of a data risk audit.
- Data Security tools – this is a book in its own right, but some approaches used can include embedding "beacons" in documents that can track the document if it gets exposed on the public web, embedding traceable rows into a dataset which can be used to

identify if a dataset has been leaked, and "exfiltration detection" that scans the open and dark web for your leaked data (CybelAngel is an example).

It's possible that you have little information available to you about these technologies. If that's the case, I recommend seeking out materials from the analyst community. The likes of Gartner, Forrester, and IDC generate plenty of material about these technologies and often rank them against various factors. You can usually find these reports online, with a little searching. Just note that a technology does not instantly become invalid if one of these major analysts doesn't feature it. For example, Gartner's Magic Quadrant may not contain tools that work well for your desired outcomes. In such cases, it's your own analysis of the tool, and its benefits, that takes precedence.

You may try to make life easier with an integrated solution from a single vendor. Or, you may engage the services of an independent consultancy to help you build out your technology stack. These are both valid options, but they also come with some key considerations. For example, certain products may only be able to retrieve metadata or manage data assets within their own technology ecosystem. Increasingly, we are seeing a move towards "best of breed" technology selections — choosing the best tool for each individual function. This works well as long as the technologies can easily integrate and "talk to" each other.

Also, be wary of taking advice from a consultancy that offers services for the technologies that they also review and recommend. A conflict of interest can arise where the consultancy recommends a technology simply because they're more familiar with it (or more likely, they have staff who know the tool well and they want to sell you their services).

Choose the Technologies That Are Right for Your Organisation

There's an enormous number of technology solutions that *might* help you in implementing your data governance policy. However, your organisation has unique goals, outcomes, and business rules. When choosing a technology, focus on how it aligns with those things. The most impressive feature list in the world doesn't mean anything if the tech doesn't serve your purpose. Remember this and you're in a better position to make good technology decisions.

CHAPTER 11

The Data Governance Roadmap

In 1988, I was living in a slum in the Philippines called Tondo. The estimates were that up to six million people lived in the shacks and shanties of that three-and-a-half square miles straddling the railway line between Caloocan to the north, the Chinese Cemetery to the east, the Divisoria street market to the south, and North Bay to the east. I was part of a community development project that ran a centre assisting families with malnourished and sick children who couldn't afford treatment through the local medical system. We treated everything from infected cuts to tuberculosis, and everything in between.

In 1989, fire swept through Tondo, burning down many of the shacks and consuming the meagre livelihoods that the locals had been able to build up over time. The disaster was exacerbated by the fact that fire trucks could not access the area to extinguish fresh spot fires as they broke out from embers floating on the hot winds. There was no roadmap showing how to get to areas with fresh plumes of smoke. As a result, the disaster was many times worse than it needed to be.

Of course, nobody ever sat there and said, "Let's build a city within a city where no emergency vehicles or first responders can get in." However, because the shanty town had sprung up little by little over time, with new arrivals adding new shacks to existing shacks, there was no map showing access points and thoroughfares.

The analogy to our data landscape is simple. Most organisations have seen databases installed and used here and there in response to tactical needs and business imperatives. Over time, these have been built on top of each other with complex interdependencies that are no longer documented or understood. No one knows the impact of making a change to some ancient database that's been serving a particular purpose well for over 20 years, and the impact of an issue with that system may not be well understood. Sometimes, a little "data paleontology" is needed to figure out what data and processes are depending on what other data and systems.

Implementing a data governance initiative is not a trivial project. In fact, I'd argue that it's not really a project at all; and approaching it like a project with an end date is one way to ensure long-term failure. Data Governance — given where most large organisations are now with their data governance (hint: see slum story above) — is a transformation. It includes transforming organisational culture as well as processes and behaviours, coupled with implementing supporting technologies, all supported by a commitment to maintaining and evolving all of those things over time.

Therefore, the successful implementation of data governance (or more correctly, the successful *improvement* of data governance — since data governance is happening in pockets all over the organisation but in a tactical, reactive and uncoordinated manner) requires a roadmap. The roadmap assesses the current state, the key stakeholders, the business value, and the organisational and strategic context of the data governance initiative. It then sets out the principles, policies and processes that define data governance in the organisation going forward, establishes the organisational structure and roles to support and enforce those policies and processes, and sets out the technology and knowledge management frameworks that are needed for sustained and efficient ongoing success. It also assesses the change management and risk management aspects of the initiative in order to maximise value and minimise disruption and confusion as the transformation is implemented.

This chapter explores these steps in detail. This is a step-by-step guide to creating your Data Governance Roadmap.

Step #1 – Data Asset Stocktake

In the first stage of creating your roadmap, you identify the key information assets that you need to manage. Consider not just the data that you already have but also data that you'd like to have, as there may be ways to obtain access to desirable external datasets as well.

Hence, the scope of your Data Asset Stocktake includes the following:

- Datasets that you own or produce internally. These are typically the results of your business processes. For example, you produce and own the dataset that an online form generates.

- Datasets that are in the public domain that you can access and leverage, such as statistics from the ABS, "open government" datasets, research datasets and so on.
- Datasets that are available commercially, either as a one-off purchase or as a subscription.

One way of approaching this stocktake is to begin by thinking about and identifying the key business entities and subject areas that concern your business, and what you want to know, analyse, or predict about each of them. This may include, for example:

- Customer (and customer lifecycle)
- Product
- Sale and sales history
- Web-click / web funnel / customer journey
- Finance data; HR data
- In insurance: Policy (and policy lifecycle), Claim and claim lifecycle
- And others that matter to your organisation's specific business model

For each key business entity, identify and map out:

1. the processes that create or acquire the data
2. the systems that store the data
3. any processes that modify the data
4. all repositories that hold records of the data
5. other datasets that enrich or could be used to enrich the data
6. downstream processes that are dependent on the data, including data warehouses / data lakes, reports, BI, analytics, predictive models
7. business processes that are dependent on the data (what human or automated decisions is the data used for)
8. the sensitivity (confidentiality) of the data, including privacy considerations

All of this information should be captured in an Information Asset Registry. This becomes foundational to your Business Glossary and your Dataset Dictionary.

Note that your Data Asset Stocktake can be significantly automated using "Data Catalog" technologies which trawl the network for data repositories, capture the technical metadata, and then use AI to categorise the data and, in some cases, to reverse-engineer the data lineage. However, these tools do not automate the thinking that is an essential part of points 1, 3, 5, 7 and 8 above.

Step #2 – Stakeholder Identification and Engagement

Given that each key data asset needs to be assigned a formal owner (per CPG235), you need to decide how to map data to owners. This could be done by assigning owners to specific production systems and their data, or by identifying key business entities (such as "Customer"), critical data elements (CDEs) or analytical datasets. Note that the owner of a business entity, CDE or dataset may change as the data progresses through its lifecycle. There might be an owner of the customer entity in the CRM, a different owner for customers in the claims system, and a different owner for the Single View of Customer (SVC or SCV) in the data warehouse.

For the most important entities, a senior executive generally needs to (be appointed to) take responsibility for the data lifecycle and data governance of data for that entity. For the customer entity, it could be the Chief Marketing Officer or Chief Customer Officer, or one of their direct reports.

In terms of the "Data Governance Organisational structure," these senior executives may have a matrix of Data Stewards who are responsible for specific datasets and/or systems and processes; for example, a data steward for CRM-hosted customer data, another one for digital marketing data and marketing automation, and another data steward for call centre customer data. (And just to throw in a curveball, the data on call centre *performance* might be managed by a different data steward looking after operational metrics and reporting to the COO!)

Step #3 – Environment Reconnaissance

This is a step that you'll apply to each individual dataset, and it comprises of two key actions:

1. Identification of all regulatory, internal, and external policies that need to govern how you use the data, who can access it, and for what purposes.
2. Identify any community expectations or ethical principles that inform how the data can be used or accessed and for what purposes.

It's important to be informed by the legislative, contractual, and ethical boundaries around the use and visibility of data. Normally, this is the remit of the legal function within an organisation (it's not the data steward's role to interpret legals and contracts) — this department will work alongside your Data Stewards in creating guidelines, with the stewards responsible for seeing that those guidelines are adhered to (for example, by working with IT on the permissions model over the datasets).

The Data Steward should take responsibility to reach out to the parts of the organisation that can provide the guidelines around data privacy and other legal, contractual, and ethical implications for the datasets they are the steward of. (Note that they may need to be sponsored into these meetings or forums by their executive-level data owner.)

Any breaches or risks that are identified during this process can be escalated through the Data Governance organisational hierarchy.

Step #4 – Value Roadmap

This is where you outline the benefits to the organisation of better data governance. It makes sense to do this specifically for each dataset, thus demonstrating practical value to senior management. This is a crucial step in terms of obtaining buy-in for your data governance initiative.

Look for opportunities against each dataset to:

- Increase revenue
- Improve business outcomes through better use of better quality data

- Accelerate time to value for data science initiatives
- Develop more accurate predictive models; better AI and process automation (RPA/IPA)
- Reduce costs and complexity
- Manage and reduce risk, including the risk of legal or ethical breaches
- Improve the quality of decision-making (human and machine)
- Reduce confusion and rework resulting from unclear definitions or conflicting BI reports and management reporting
- Improve employee satisfaction and engagement due to empowerment and confidence in data and decision-making
- Data monetisation by commercialising data for other parties who might pay for access to your datasets
- Data philanthropy, where you provide certain datasets for free to the wider community who might drive social development projects or analyse trends to inform better policy and government decisions.

This process, in turn, supports the prioritisation of subject areas to address, based on expected business benefit from improved governance (data quality, availability, security, etc.) and risk management of the dataset(s).

Step #5 – Cross-Inform the Information Management (IM) Strategy

Your Information Management strategy is the overarching strategy under which your data governance roadmap fits. It may exist in your organisation as an artefact with the IT department, or the CDO may be charged with developing or renewing the IM Strategy.

The Information Management strategy typically considers the following areas:

	Current State	Target State	Transition Plan
Information Value Roadmap	Re-baseline business requirements for information value creation and access	Design future state based on refreshed information and access requirements	Prioritised Roadmap for Business Value from Analytics, AI, ML and RPA.*
Systems & Technology	Review of current technology ecosystem against business capability requirements	Recommended technology ecosystem and architecture to meet business needs	Alignment of technology releases with information value roadmap
Processes and Methodology	Assessment of current capabilities and processes	Sustainable approach to capability development and data governance	Incremental approach to implement new processes, methodology and governance
Organisation and Skillsets	Review of current organisational structure and capabilities	Required structure, roles and skillsets to support target state capabilities, processes and governance	Resource gap and transition plan aligned with information and technology roadmaps

There may or may not be a broader Information Management Strategy in your organisation which overarches the Data Governance Roadmap. If there is none, that's fine, we push on. If there is, they just need to be in sync.

Step #6 – Principles, Policies, and Processes

Your Data Governance Principles are the values that will guide decisions related to the scope of the data governance initiative and how it's implemented. These principles inform the data governance policies that you create, and these policies, in turn, guide the development of the processes used in the organisation.

Just like your organisational vision statement and values, your data governance principles need to be socialised across the organisation and broadly agreed with, so that they can be referred to as necessary when decisions need to be made. ***This is how you initiate the cultural change that enables your organisation to become "data-driven" and to place the appropriate value on looking after data properly. This is how you empower your organisation to reap the benefits of the Fourth Industrial Revolution.***

There will be several rounds of stakeholder engagement as you socialise these principles and help the organisation and its leadership to unpack and embrace the implications.

At a minimum, these rounds would include:

1) Defining the data governance principles in line with the organisational values and getting agreement and formal

endorsement at the most senior levels (the board and the C-suite) that this is how the organisation will treat its data and related processes. (If the board have not yet bought into data governance, the CDO might need stronger sponsorship to get the appropriate attention at that level — or, it may be time to consider whether the board is able to take the organisation where it needs to go in terms of the competitive environment and laying the right foundations for AI, data science, process automation, and the next industrial revolution.) Once this endorsement is obtained, you need a round of internal "advertising" to inform everyone who works with data that these data governance principles have been endorsed by senior leadership and will now guide the formulation of data policies and processes that will positively impact the culture of the organisation and drive the organisation to greater productivity and competitive advantage as a data-enabled, "Fourth Industrial Revolution" organisation.

2) With the key stakeholders identified in Step 2, drill down on the agreed values and principles to derive appropriate, detailed policies based on these principles. What these policies look like might vary to some extent between different parts of the organisation, which is why it's important to involve a broad representation of senior data stakeholders.

3) Each Data Executive or Business Data Champion works with their data stewards to define the processes that are needed to implement the principles and policies.

I've provided a couple of examples of what Data Governance Principles might look like:

Example 1:

Principle	**Description**
Data as an Asset	Data will be regarded (thought of) and managed as a corporate asset, giving rise to similar processes as for value-based asset management.
Fit for Purpose	Before data is employed for a new function or purpose, the data will be evaluated for fitness against that purpose, using the key dimensions of the Data Governance Framework (Usability, Security/Availability, Quality).
Business-Driven	Data remediation initiatives will have a business driver and business outcome (we don't "fix for the sake of fixing").
Ethics	Before data is employed for a new function or purpose, a legal (contractual and legislative) and ethical check will be undertaken and signed off.
Sharing	Except where subject to privacy or ethical considerations, the default position is that data is open to be shared with any teams that can derive value from the data.

Example 2:

Principle	Description
Shareholder Value	Data will be treated across its lifecycle in a way that maximises the value returned to our shareholders.
(or) Triple Bottom Line	Data will be treated across its lifecycle in a way that maximises the value returned to our stakeholders including employees, shareholders, the community, and the environment.
Efficiency	Data will be available to those who need it at the right time, with the right context and understanding (caveats) and in the right format.
Accountability	Data that is "governed" will be overseen by the business area (and its manager in their role as such) that has the appropriate knowledge, authority, and accountability to define how the data is managed, protected, and shared within and outside of the organisation.
KPIs	Metrics for success in data management will be established and tracked as part of the KPIs of the business area that is overseeing the data (see Accountability above).

As the Principles are discussed and defined, the following points will probably be considered and should be captured as the detailed version of the principles:

- Principle title (as above)
- Short description (as above)
- Long description — clarifies any ambiguities or shades of meaning

- Rationale/justification — why is this principle important?
- Benefits — what positive outcomes are expected from adherence to this principle?
- Implications — this is where you begin to unpack the Policies that are implied by the Principles!

The data governance principles, the policies that have been derived from them, and the processes that result from those policies, need to be retained in a format that is accessible by the data stewards as needed. They will need to be referred to in the milieu of day-to-day data management as they provide the "why" as well as the "how" to all the policies and processes that people can easily grow tired of… (which would begin the pendulum swing back towards data chaos).

Step #7 – Organisational Structure, Bodies, and Roles

There are numerous organisational models that you can implement at this stage of the process. I believe the following is one of the most effective:

Here, we have three organisational layers.

At the top, we have our Data Governance Executive. The executive consists of business leaders at the C, EVP/EGM or VP/GM level, who have executive ownership of one or more of the key datasets defined in step 1. This is an additional role to their "day job" responsibilities. The executive acts as an escalation and resolution point for data-related issues that cannot be resolved by the Business Data Champion. They may do this as part of a formal monthly forum, or more reactively and informally, via meetings as needed. However, note that trying to raise and solve issues at monthly committee meetings is going to introduce a slower way of working; more agile ways of getting these decisions made and endorsed might be worth exploring.

Each of the data executives will manage one or more Business Data Champions. In large organisations, the Business Data Champion is a full-time role within a particular line of business and/or related to the specific key datasets identified in Step 1. The Business Data Champion's area of responsibility aligns with that of the executive owner of the same datasets or critical data elements (CDEs).

The Business Data Champion also:

- Acts as the escalation point for Data Stewards where conflicts or issues cannot be resolved at the Data Steward's level.
- Acts as a mentor and coach to multiple Data Stewards.
- Represents their subject areas and data repositories (per their data stewards) on project steering committees, data governance forums, and the like.

The Business Data Champion typically reports to the CDO, with a 'dotted line' to the Data Governance Executive who is responsible for the corresponding datasets and data repositories, as explained above.

Finally, we have our Data Stewards, whose role we've already discussed. These are the people at the coal face who ensure that their part of the organisation adheres to the data governance principles, policies, and processes. As with the Data Executive, this is often a part-time role in addition to the person's day-to-day responsibilities as an analyst, subject matter expert or information worker.

In my Data Governance Masterclass, I highlight the following aspects of the Data Steward's role:

- A set of additional responsibilities awarded to a person within a business unit.

- Part influencer, part knowledge curator; with responsibilities to advocate for improved data governance and data quality in BAU activities and in projects impacting their area.
- Acts as the SME on data-related requirements, issues, and roadmap going forward for the datasets and data repositories within their purview.
- Helps define the roadmap for how data and the data strategy can and will support the business function(s) they are aligned with.
- Identifies opportunities for data quality uplift in BAU processes and data management process improvement.
- Identifies opportunities for data governance and data quality uplift in projects and enhancements that touch on the data repositories within their remit.

Step #8 – Permissions, Access, and Security

For each dataset in the Data Asset Stocktake, you need to determine its sensitivity based on your findings during the Environment Reconnaissance step. Once you've done that, there are a couple of ways you can approach a security and permissions audit.

The first is to determine who currently has access to each dataset and to what level of detail. The use of a Data Catalog technology can significantly accelerate this step, as most of this scanning should be able to be automated. From this, determine if there are any breaches in security and permissions for any dataset that need to be rectified. Once that's done, define a process whereby access to a dataset can be requested, reviewed, and approved or denied/modified.

The second approach is to implement a security layer that reflects the information security requirements. You then wait for a while to see who in the organisation complains about getting impacted. From there, you can make changes to the security later on a case-by-case basis.

Step #9 – Knowledge Management

In case you haven't picked this up already, knowledge management is a huge component of data governance.

Knowledge management for data governance requires three components to be successful:

1. A training and education program (ideally for everyone but, at a minimum, for the data stewards) on the importance of knowledge management, which will also create motivation to proactively manage data knowledge as a corporate asset. The data stewards can then champion these values across their stakeholders.
    - Note that this can be very effectively done as part of an organisation-wide "data literacy" training program.
2. Technology to support knowledge capture, curation, and dissemination / availability. This could begin with your data catalog's automatic capture of metadata and data lineage, which can then be enriched by data stewards and other information workers and knowledge workers improving and enhancing those machine-generated elements over time. This continuous improvement of dataset knowledge needs to become part of the corporate culture (which you would inculcate partly through the training mentioned above). You need a clear and user-friendly ongoing process to capture more and more tacit knowledge over time. (Ideally, people are thinking "I've got some free time, I'll jot down that thing we discovered last week about the web click dataset.")
3. Awareness for ***all*** information workers on how to access the knowledge that is available — hence you need to promote the Business Glossary, Dataset Dictionary / Data Catalog, Report Inventory, and any other knowledge repositories, so that people know how to easily access the information that can help them do their job more efficiently, effectively, accurately, and confidently.

Step #10 – The Technology Ecosystem

I've provided an entire chapter to the role that technology can play in your data governance initiative. Remember the key takeaway — technology is a tool rather than a solution.

Obviously, you'll need to collaborate with IT to determine which technologies offer the most value, including some that might already have been purchased. The right people in IT will be able to help with templates and approaches for functional and non-functional requirements, evaluation criteria and so on, while Procurement generally looks after negotiating with potential suppliers.

Step #11 – Change Management

Implementing a data governance policy often involves making changes to roles and responsibilities and introducing new processes and technology; change can be upsetting and is often passively or actively resisted. It's best to engage a professional change management resource to provide guidance across the whole data governance initiative. This may be a person who's part of your internal project management office (PMO) or a specialist consultant that you engage on contract.

Aspects of change that need to be considered include the following:

- Cultural change around the value of data, the management of data risk, levels of data sharing, the capture of data knowledge, and the formalisation of previously informal processes.
- Communication of the changes to impacted stakeholders, to maximise buy-in and minimise resistance.
- New roles requiring training; for example, "So now I'm a part-time Data Steward – what does that mean?"
- New organisational bodies such as the Data Executive team or a Data Governance Group, Reference Data Councils, and so on.
- Changes to existing processes, and entirely new processes (e.g. data risk identification, escalation, and resolution)
- Implementation of Knowledge Management processes
- Reduced levels of access (e.g. some people may no longer have access to PII)
- New technologies requiring training

The final aspect to consider when creating your data governance roadmap involves managing the risks it may entail. You must consider both external and internal threats here.

On the external side, consider who could benefit from having improper access to your data. What methods might they use to gain access and what can you do to counteract them. It's often best to employ cybersecurity experts here.

On the internal side, consider the possibility of a disgruntled worker using data improperly. What measures can you take to prevent this from happening and how will you respond if it does?

Finally, consider the risks that may be introduced by the changes you're making. There are risks involved in making any changes, so change risk management needs to be a part of your risk management. Additionally, consider what new risks might be introduced once you've implemented the changes you need to make.

Of course, you may not have the authority to make decisions about every aspect of your organisation's data governance practices and policies. However, you can play a critical and effective role in creating and maintaining a thriving data culture in your organisation regardless of your place in the hierarchy or the specificity of your formal job scope.

How Does Data Governance Apply to Your Role?

As an information worker in your organisation, the specific answer will vary depending on the organisation and your role within it. Typically, data governance applies most practically to the following areas:

- **Knowledge Management** – This relates to managing what you know about the data that you collect. For example, you need to understand the source of the data and what's happened to the data on its journey from where it was first collected to where it's now being accessed. You also need to know who has access to the data, as this may be a question that regulatory authorities will ask if your organisation deals with sensitive data.

- **Risk Management** – Data governance often involves the identification of risks that may arise from using the data. These may also relate to regulatory and contractual issues, in addition to potential ethical and security breaches. This aspect of data governance also relates to the processes that you have in place should a breach occur. Again, I refer back to the email example I've shared previously. The lack of risk management in place meant that a BCC becoming a CC created a six-figure problem for a company.
- **Change Management** – Whenever you try to change how people work, there is always an aspect of data governance. You're usually showing people how to use a new system, which means they have to understand how to use said system correctly. In some cases, this may involve a full-blown cultural change centred around the importance of data.
- **Change Risk Management** – Introducing change creates an element of risk. And that risk grows as the change becomes more complex. In many cases, your role is to mitigate this change-related risk to ensure no data breaches occur during the change process.
- **Issue Management** – How do you handle data-related issues within your organisation? It's this question that leads to the creation of processes for every data-related issue you may face.
- **Environmental Awareness** – If you work in a fast-paced and ever-changing environment, environmental awareness becomes a key part of data governance. You need to ensure that relevant areas of the business have access to key data. Plus, you need to get ahead of the issues that can arise because of this fast pace. For example, data governance in this area may involve staying ahead of emerging cybersecurity issues.
- **Strategy** – Perhaps you're responsible for creating the entire data governance strategy that your organisation relies upon. A good strategy supports and integrates with your analytics and data science strategy. It may also prepare you for the use of new technologies, such as AI and ML.
- **Organisational Bodies** – These are the groups that create the rules that govern the use of the data in your business. They often use industry guidelines to determine which data is sensitive and which

isn't. They may also need to apply degrees of sensitivity, depending on the data. The key function of an organisational body lies in ensuring that data doesn't get into the wrong hands. This means it's responsible for all of the checks and measures within your data governance strategy. Plus, it's the organisational body that builds accountability into the strategy.

- **Enforcement Methods** – Your role may involve figuring out what the organisation does if a breach occurs. In creating enforcement methods, you're developing a set of steps to follow when something unexpected happens. You may also need to prioritise work and projects to ensure maximum impact and benefits from your data.

Follow the Roadmap to Success

You may notice that the steps for this roadmap are essentially a round-up of the topics that I've covered in the book. It brings everything together into a cohesive whole that leads to the creation of an effective data governance strategy and roadmap.

Don't get stuck without a roadmap like the firefighters in Tondo.

Develop your roadmap and revisit it often. Use it to identify and navigate the key challenges you'll face when building data into the culture of your organisation.

Sharpening the Saw:
The Data Governance Health Check

Industry best practices are constantly evolving, as is the environment that business (and data governance) takes place in.

A Data Governance Health Check or review by an external party every two or three years can greatly assist in improving the effectiveness and efficiency of your data governance practices, bringing a fresh set of eyes to your ways of working and introducing new best practices from the industry as it evolves. This kind of regular review can also be used to demonstrate to regulators your organisation's commitment to best practice and ongoing data governance improvement.

At the time of writing, I'm halfway through a Data Governance Maturity Assessment and Roadmap for a large insurer. This company is well into their data governance journey but sought out some expert advice to create a benchmark from which to measure progress and to fine-tune the organisational model they need to support it.

How I Can Help

Through the company I lead, I can provide a number of services to help your organisation move to the next level of data governance maturity:

Data Governance Maturity Assessment – this is a six-week engagement (typically) that assesses the current state of data governance maturity across 18 dimensions. It provides a set of recommendations for each, and then blends these recommendations into a cohesive roadmap for presentation back to an executive audience.

Data Governance Training – a 2-day masterclass for data stewards, Business Data Champions, and interested executives. It provides the opportunity to deep dive into the contents covered in this book and discuss ideas and implementations in workshop-style activities with your peers.

Data Governance Coaching – sometimes you don't need a big bang, but you do need a second set of eyes to provide executive coaching specifically to assist in the successful execution of a data governance roadmap. This is a 12-month engagement with meetings every two weeks, following an "executive coaching" format specifically for data governance.

Find me on LinkedIn and let's have a conversation about whether any of these services could be of value to your organisation.

SECTION 4

The Future of Data Governance

CHAPTER 12

What Does the Future Look Like for Data Governance?

As I discussed in the first few chapters, the Fourth Industrial Revolution will enable huge strides in increasing human productivity, in automating the dull, dangerous and dirty jobs, and in driving innovation and value creation in ways we are just beginning to explore.

There are certain aspects of a Data Steward's role that would qualify as "dull" (let's be honest) and indeed we are already seeing AI and machine learning begin to impact some of those tasks. The automated collection of metadata into a data catalog is one example, as the software interrogates every data repository for its technical metadata, and then uses AI to derive some of the characteristics of the dataset. Automated data profiling to create a data quality score and benchmark is another. We are also beginning to see AI used to "reverse-engineer" data lineage by stringing together the paths between source and target data repositories. And emerging technologies like Okera can draw conclusions as to data sensitivity (such as personally identifiable information) and automatically apply access and security rules over every instance where the data is found.

All of this makes the Data Steward's job and, ultimately, the CDO's job much easier and faster.

Of course, this brings us to the key question that anybody has when we raise the issue of AI — what (or perhaps who) will become redundant?

What I can say in response to this is that I don't see AI making the role of the Data Steward redundant. Instead, it will eventually automate whatever can be automated and make the role easier and the steward more productive. A useful guideline is, "If you can document it, you can automate it." That is, if the logic is able to be explained and captured in words, no matter how complex or nuanced, then an AI will be able to automate that process. Humans will manage the exceptions, the edge cases, and the situations where the AI can't make a decision within the bounds of "confidence" that it has been provided with.

New Levels, New Devils

The prevalence of AI, of course, brings its own challenges, and just as we have discussed the need for legislation, community expectations and ethical considerations to govern the use of data, so too will these considerations need to be taken into account when governing the use of "big data", pervasive sensors, IoT, artificial intelligence and machine learning.

Human history shows that people and parties with power tend to accumulate more of it and tend to not let go without being forced. The story of the Magna Carta is an example of how almost unlimited power (vested in the king) was brought under the control and accountability of a wider group of people representing the population at large. This historic watershed was in fact an anomaly in the overall trend of human history but is the foundation of the freedoms we enjoy today. Democracy is a fragile balance of vested interests, processes, and power plays; governments empowered by AI and pervasive monitoring tools will be ever tempted to exploit them to reduce transparency and accountability, to ostracise or eliminate (perhaps by simply deplatforming) critics, and to perpetuate their entrenched positions and privileges. We see this already with tech-enabled dictatorships across the globe; as I write, this is particularly pernicious with the Chinese communist government. The same will be true of some large multinational corporations censoring free speech in a completely unaccountable way simply because they can.

In all seriousness, power — whether in the form of pervasive big tech or invasive government intrusion — cannot be left unchecked. The legal framework for data ethics to keep both governments and corporations accountable will need to keep a step ahead of the behaviours of powerful vested interests if we are to avoid a scenario reminiscent of George Orwell's dystopian *1984*.

For Organisations

One thing that we can say for certain is that the data revolution will not slow down. In fact, we're only at the beginning stage of something that will change everything about how we do business and operate in the world. For, as important as data is now, it will become many times more important in the future.

In fact, I would argue that we are at the beginning of the end for organisations (outside of small businesses and, perhaps, agriculture) that do not have a strong data governance practice. Without this, you're entering into a new age without the tools that you need to succeed in it. And as we've seen throughout history, organisations that can't adapt to disruption get left behind.

In fact, we're likely to see corporations fall into one of three categories in terms of their response to the opportunities of the Fourth Industrial Revolution.

There will be companies who continue to do business the way they always have and assume that their business model is strong enough to last. These companies will go the way of Kodak, Blockbuster video stores, record companies bankrupted by the Spotify business model, and so on. They will not survive.

Other companies will see the writing on the wall and will seek to rapidly adopt AI, machine learning, and process automation and will set ambitious goals of becoming data-driven or analytics-enabled organisations. However, having never paid much attention to data governance, the data they seek to help drive them will be exposed as not fit for purpose, and they will stumble at the starting blocks of their journey towards automation and value. Then, they will lose a year or two as they retrospectively try to get their data house in order — precious months during which their competition will leap to fill the gaps in the market that these companies have left open. Furthermore, not being familiar with the disciplines and value of data governance, they are likely to take the same shortcuts over AI and model governance, meaning that their second leap from the starting blocks will also send them crashing in a heap of broken machine learning models and ethical AI dilemmas.

Finally, there will be companies who already have their data "house" in order and understand which datasets are fit for the purposes of AI, ML, and process automation. These companies will know which datasets can support the automation of which business processes and will be able to more quickly and fully benefit from the automation and acceleration of the Fourth Industrial Revolution. They will define their markets while their competitors flounder.

Which kind of organisation will yours be?

Your Next Steps

I'd like to take this opportunity to thank you for journeying with me through all of the areas covered in this book. I hope that I've succeeded in helping you to understand why data governance is more important now than ever before. And I hope that the viewpoints, frameworks and processes I've shared can help you to begin the transformation process in your organisation.

There are two hurdles that an organisation needs to overcome if it's going to implement good data governance practices.

The first is recognising that something needs to change and finding the right catalyst to engage executive sponsorship for that change. If you're in an organisation right now where the need for growth in this area is not getting traction in the C-suite, you may have to find the right way to frame the imperative. Good data governance supports cost savings in IT development activities, it offers competitive advantage, it supports innovation, it reduces risk in a number of ways, and it sets up your organisation for success in the data economy. We looked at a number of these and more benefits in Chapter 3. Which of these benefits is the "theme" that's getting traction with the C-suite at the moment? Sometimes, there's a huge focus on risk because of some other organisation getting hammered by the regulator or in the media for poor practices; if that's the case, you may get better sponsorship for a data governance uplift by playing the risk mitigation card. At other times, it might be "innovation" that carries the magic aura that gets attention and sponsorship. Sometimes, positioning what you're offering in terms of what already has traction can get you further.

Once you've overcome the hurdle of "why", we engage the hurdle of "how". I recently completed a consulting engagement with a large organisation (over 20,000 staff) where, following a comprehensive (but rapid) consultation process with key stakeholders, we created a data governance roadmap for the next stage of their journey. Typically, an engagement like this can be completed in a matter of weeks and produce a roadmap that provides direction and value over one to three years. Find me on LinkedIn and let's have a conversation about whether a service like this could be of value to your organisation.

I can also provide training courses for the data stewards and Business Data Champions to help get them oriented to the role they will be required to play in the context of data governance. I also have a network that can help find the right data governance professionals to bring into your organisation if needed.

I look forward to helping you further in any way that I can.

Made in the USA
Monee, IL
04 November 2026

67383107R00092